QUANTUM GROUPS AND NON-COMMUTATIVE GEOMETRY

Les Publications CRM
Directeur: Francis Clarke

Lecture notes/Notes de cours:

Quantum Groups and Non-Commutative Geometry
Yu. I. Manin, 100 p., Course given at CRM, 1989.

Une Introduction à la Théorie des Équations aux Dérivées partielles
S. Zaidman, 138p., Univ. de Montréal, 1989.

Representations of Real Reductive Groups
A.W. Knapp, 257 p., CRM Summer School Course, 1990.

Introduction à la Topologie des Ensembles Fractals
Cours de C. Tricot donné à l'École Polytechnique, Notes de cours de S. Baldo, 80 p., 1991.

Groupes Linéaires Algébriques
R. Bédard, Cours d'été CRM, 1991.

Asymptotic Theory for Bootstrap Methods in Statistics
Edited by G. Ducharme and B. Boulerice, Course given by R. Beran at CRM.

A Survey of the Hodge Conjecture
J.D. Lewis, CRM Summer School Course.

Proceedings/Actes:

Analyse Non Linéaire (co-édition Gauthier-Villars)
Éditeurs: H. Attouch, F.H. Clarke, J.-P. Aubin et I. Ekeland, 482 p.,
Colloque France-Québec de Perpignan, 1989.

Hamiltonian Systems, Transformation Groups and Spectral Transform Methods
Editors: J. Harnad and J.E. Marsden, 247 p.,Proceedings of a CRM Workshop, 1990.

Lie Theory, Differential Equations and Representation Theory
Editor: V. Hussin, 464 p., Proceedings of the CMS Annual Seminar held at CRM, 1990.

Automorphic Forms and Analytic Number Theory
Editor: Ram Murty, 150 p., Proceedings of a CRM Workshop, 1990.

The Zeta-Functions of Picard Modular Surfaces
Edited by R. Langlands and D. Ramakrishnan, Proceedings of a CRM Seminar.

Monographs/Monographies:

Optimization and Nonsmooth Analysis
F.H. Clarke, 310 p., 2nd edition 1989.

Tables of Representations of Simple Lie Algebras
Vol. 1: Exceptional Simple Lie Algebras
W.G. McKay, J. Patera and D.W. Rand, 350 p., 1990.

SimpLie™ User's Manual and Software
W.G. McKay, J. Patera and D.W. Rand, 50 p.+ Software, 1990.

ISBN 2-921120-00-3

Dépôt légal 4ième trimestre 1988, réimpression 1991

Centre de Recherches Mathématiques
Université de Montréal
C.P. 6128, Succursale A, Montréal, Québec, Canada, H3C 3J7

Quantum Groups and Non-Commutative Geometry

Yu. I. Manin

Centre de Recherches Mathématiques (CRM)
Université de Montréal
C.P. 6128 - A
Montréal, QC H3C 3J7
Canada

CONTENTS

Page

INTRODUCTION

We begin with some terminology and background, in particular we define Hopf algebra and quantum group.

Let H be a function algebra of a Lie group G. Then the multiplication map $G \times G \to G$ (resp. the inversion map $G \to G : g \to g^{-1}$, resp. the inclusion of the identity point) gives rise to a "comultiplication" $\Delta : H \to H \otimes H$ (resp. "antipode" $i : H \to H$, resp. "counit" $\varepsilon : H \to \mathbf{R}$) which satisfies a set of identities defining the general algebraic notion of a Hopf algebra (cf. §2 below). This set is symmetric with respect to reversing all arrows. However, an asymmetry does arise: while H as an algebra always has a commutative multiplication it may well have a non–commutative comultiplication. (This is the case if G is non–abelian.)

Hopf algebras that are not necessarily commutative or cocommutative have been studied by algebraists for several decades (cf. Abe[A]). Recently, however, some very specific Hopf algebras emerged in mathematical physics and were christened "quantum groups". Initially, they appeared in the quantum inverse scattering transform method (QIST) developed by L.D. Faddeev and his school (cf. [STF], [FT], [F], [S], [FRT]) and reviewed in V.G. Drinfeld's Berkeley talk [D1]. Closely related work (also motivated by QIST) was done by M. Jimbo [J1], [J2] and, from a somewhat different viewpoint, by S.L. Woronowicz [W1], [W2]. One idea behind this work is that such rigid objects as classical simple groups (or Lie algebras) in fact admit continuous deformations as Hopf algebras and the deformed objects close to the initial one can be described very precisely together with their representation theory.

In this work, we systematically develop a different approach to quantum groups, based upon the following observation. Suppose that you "quantize" the simplest phase plane, imposing on its coordinates the commutation relation $xy = e^{\hbar} yx$; i.e. the <u>integrated</u> version of the Heisenberg commutation relations. Then the ordinary symmetry group GL(2) of the plane breaks down. However, this "broken symmetry" is completely restored if one imposes some nontrivial commutation relations upon the entries of the 2×2–matrices. Thus, one arrives at the notion of the quantum group $GL_q(2)$, $q = e^{\hbar}$, which is described in complete detail in §1.

One remarkable property of this approach is its generality. Namely, instead of a "quantum plane", in §§3,4, we start with a "quantum linear space" defined by **arbitrary quadratic relations** among its noncommutative coordinates and we obtain its "general linear quantum group", or rather a pair: "quantum semigroup of endomorphisms" $\rightarrow$ "quantum group". The first object is a noncommutative space of the same kind (i.e. it is defined by quadratic relations) while the second one is obtained from it by a process of noncommutative localization as advocated in algebra by P.M. Cohn, and arising here for the first time in a natural way. The point is that one inverts matrices and not just elements of a ring, and to obtain a Hopf algebra, one generally must invert infinitely many matrices.

We use the word "noncommutative space" in the spirit of Alain Connes. The main difference is that we develop a chapter of noncommutative algebraic geometry while Connes deals with differential geometry and topology. In §10, we discuss a way to introduce a $*$–structure in our groups thus making it possible to define their compact forms.

As in [W1], [W2], [FRT] but unlike [D1], we work with the noncommutative function ring of a quantum group rather than its "universal enveloping algebra", which is a dual object. Of course, both deserve close attention, but in our approach the former appears more naturally. The key technical notion in this connection is that of a multiplicative matrix: cf. §2.6–2.10. It is also worth mentioning that we have no need to consider only small deformations of classical objects: the parameter spaces of our objects are defined globally. They are just grassmannian spaces of quadratic relations: cf. §3.2. The price paid for this is that we lose the notion of a "semisimple" quantum group (which anyway has never been formalized in previous work).

Moreover, we do not need to impose any relation to the Yang–Baxter equations from the start. However, we can and must explain the relation at a later stage. An important thing to remember is that a Yang--Baxter operator generates quantum groups by two very different procedures. One is to consider the quantum automorphism group of a quadratic algebra which is just a "Yang–Baxter symmetric algebra". The groups appearing in QIST are of this type.

Another way starts with a relativization of the notions of quadratic algebra, quantum group etc. by replacing everywhere the transposition of factors in a tensor product by a Yang–Baxter operator. The simplest example is the "super version" of our constructions, which is fairly obvious. Other examples were not considered before [M]. We explain these ideas rather succintly in §§11,12.

An earlier version of this work is [M]. Many details and some new results are added here. I would like to mention in particular the general construction of a Hopf algebra starting with a bialgebra with a generating multiplicative matrix (see §7). These notes are not meant to be a survey of this quickly growing subject; the bibliography is very incomplete. The ideas of this paper were first developed in my lectures at Moscow University in the winter of 1986-1987 and at the subsequent seminar where Yu. Kobyzev suggested the description of $GL_q(2)$ which was seminal for all that follows.

These notes were written for a series of lectures given at the Centre de recherches mathématiques at the Université de Montréal in June 1988, where I was invited as Aisenstadt Professor. I am grateful to many people who made my stay in Montréal very agreeable and productive, in particular to André Aisenstadt, whom it was my pleasure to meet. I would like to thank Luis Alvárez-Gaumé and John Harnad for their assistance in the preparation of these notes. Finally I would like to thank Nathalie Brunet, Louise Letendre and Angèle Patenaude for their careful typing of the manuscript.

1. THE QUANTUM GROUP GL(2)

1. Noncommutative spaces: function rings and points. In this paper, we fix once and for all a field k. A ring (or an algebra) means an associative k–algebra with identity, not necessarily commutative. It is suggestive to imagine a ring A as a ring of (polynomial) functions on a space which is an object of noncommutative, or quantum, geometry. Morphisms of spaces correspond to ring homomorphisms in the reverse direction. For A, B fixed, the set $\mathrm{Hom}_{k\text{–alg}}(A, B)$ is also called the set of B–points of the space (defined by) A.

There is no harm in giving a formal definition of the category of noncommutative spaces as $(k\text{–Alg})^{op}$ as long as one remembers that some of the most common categorical prejudices could be misleading in $(k\text{–Alg})^{op}$. To quote just one, the tensor product in k–Alg does not define a direct product in $(k\text{–Alg})^{op}$, but morally does correspond to a "direct product" of quantum spaces.

2. Two quantum planes and quantum matrices. Fix $q \in k$, $q \neq 0$. The quantum plane $A_q^{2|0}$ is defined by the ring

$$A_q^{2|0} = k\langle x, y\rangle / (xy - q^{-1}yx) \tag{1}$$

where $k\langle x_1, \ldots, x_n\rangle$ means an associative k–algebra freely generated by $x_1, \ldots, x_n$. This is a deformation of the usual plane corresponding to $q = 1$. We shall also need a deformation of the $0|2$–dimensional "plane" of supergeometry:

$$A_q^{2|0} = k\langle \xi, \eta\rangle / (\xi^2, \eta^2, \xi\eta + q\eta\xi). \tag{2}$$

For $q = 1$, this superplane is a Grassmann algebra in two variables. Both rings (1) and (2) are naturally graded (the generators being of degree one), and the dimensions of their homogeneous components coincide with those of the symmetric and exterior algebras respectively. In fact, the monomials $x^a y^b$, $0 \le a, b < \infty$ (resp. the monomials $\xi^a \eta^b$, $0 \le a, b \le 1$) form a k–basis of (1) (resp. of (2)).

Finally, the coordinate ring of the manifold of quantum two–by–two matrices $\begin{pmatrix} a & b \\ c & d \end{pmatrix}$ is defined by

$$M_q(2) = k\langle a, b, c, d\rangle /,$$
$$(ab = q^{-1}ba,\ ac = q^{-1}ca,\ cd = q^{-1}dc,\ bd = q^{-1}db, \qquad (3)$$
$$bc = cb,\ ad - da = (q^{-1} - q)bc.)$$

Although the commutation relations (3) are slightly more complicated than (1), one easily proves that the monomials $a^\alpha b^\beta c^\gamma d^\delta$ still form a basis of $M_q(2)$.

In order to state our main theorem on $M_q(2)$, we need the following notion. Two families of elements S, T of a ring A are called commuting, if $[s, t] = 0$ for any $s \in S$, $t \in T$. An A–point of $A_q^{2|0}$, $A_q^{0|2}$, $M_q(2)$, etc. is considered as a family of its coordinates, e.g. an A–point of $M_q(2)$ is a quadruple $(a, b, c, d) \in A^4$ verifying (3). In this sense, we can say that two points are commuting. Finally, a B–point of a quantum space A is called a generic point, if the corresponding morphism $A \to B$ is an injection.

3. Theorem.

a) Let $\begin{pmatrix} a & b \\ c & d \end{pmatrix}$, $\begin{pmatrix} a' & b' \\ c' & d' \end{pmatrix}$ be two commuting A–points of $M_q(2)$. Then $\begin{pmatrix} a & b \\ c & d \end{pmatrix}\begin{pmatrix} a' & b' \\ c' & d' \end{pmatrix}$ is a point of $M_q(2)$.

b) With the same assumptions, let

$$\mathrm{DET}_q\begin{pmatrix} a & b \\ c & d \end{pmatrix} = ad - q^{-1}bc = da - qcb.$$

Then

$$\mathrm{DET}_q\left[\begin{pmatrix} a & b \\ c & d \end{pmatrix}\begin{pmatrix} a' & b' \\ c' & d' \end{pmatrix}\right] = \mathrm{DET}_q\begin{pmatrix} a & b \\ c & d \end{pmatrix}\mathrm{DET}_q\begin{pmatrix} a' & b' \\ c' & d' \end{pmatrix}, \qquad (4)$$

and $\mathrm{DET}_q\begin{pmatrix} a & b \\ c & d \end{pmatrix}$ commutes with a, b, c, d.

c) Assume in addition that $\mathrm{DET}_q\begin{pmatrix} a & b \\ c & d \end{pmatrix}$ is invertible in A. Then

$$\begin{pmatrix} a & b \\ c & d \end{pmatrix}^{-1} = \mathrm{DET}_q\begin{pmatrix} a & b \\ c & d \end{pmatrix}^{-1} \cdot \begin{pmatrix} d & -qb \\ -q^{-1}c & a \end{pmatrix} \qquad (5)$$

is an A–point of $M_{q^{-1}}(2)$. ■

In principle, all these statements can be checked directly, e.g. the last statement is quite straightforward. However, checking the commutation relations (3) for the product of two commuting points is cumbersome and not very illuminating. One does not readily see what is so special about (3).

A proper way to think about (3) was suggested by Yu. Kobyzev. After all, matrices act upon coordinate spaces, and matrix multiplication is just a reflection of this action. In fact, the same happens in the noncommutative realm.

4. Theorem. Let (x, y) (resp. (ξ, η)) be a generic A–point of $A_q^{2|0}$ (resp. $A_q^{0|2}$). Let $(a, b, c, d) \in A^4$ commute with (x, y, ξ, η). Write

$$\begin{pmatrix} x' \\ y' \end{pmatrix} = \begin{pmatrix} a & b \\ c & d \end{pmatrix}, \quad \begin{pmatrix} x'' \\ y'' \end{pmatrix} = \begin{pmatrix} a & c \\ b & d \end{pmatrix}, \quad \begin{pmatrix} \xi' \\ \eta' \end{pmatrix} = \begin{pmatrix} a & b \\ c & d \end{pmatrix}\begin{pmatrix} \xi \\ \eta \end{pmatrix}.$$

If $q^2 \neq -1$, the following conditions are equivalent:

(i) (x', y') and (x'', y'') are points of $A_q^{2|0}$;

(ii) (x', y') is a point of $A_q^{2|0}$, (ξ', η') is a point of $A_q^{0|2}$;

(iii) (a, b, c, d) is a point of $M_q(2)$.

For $q^2 = -1$, we have only (iii) $\Rightarrow$ (i) and (iii) $\Rightarrow$ (ii).

Proof. Let us check, e.g., that (i) $\Leftrightarrow$ (iii). The relation $x'y' = q^{-1}y'x'$ means

$$(ax+by)(cx+dy) = q^{-1}(cx+dy)(ax+by).$$

Since (x, y) is a generic point and a, b, c, d commute with x, y, this equality is equivalent to the equality of the three coefficients:

$$\begin{aligned} x^2 &: \ ac = q^{-1}ca, \\ y^2 &: \ bd = q^{-1}db, \\ xy &: \ ad - da = q^{-1}cb - qbc. \end{aligned} \tag{6'}$$

Exchanging b and c, we get the relations equivalent to $x''y'' = q^{-1}y''x''$:

$$ab = q^{-1}ba, \ cd = q^{-1}dc, \ ad - da = q^{-1}bc - qcb. \tag{6''}$$

Comparing the last relations in (6′) and (6″), we obtain

$$(q+q^{-1})(bc-cb) = 0 \ \Rightarrow \ bc = cb, \text{ if } q^2 \neq -1.$$

Hence (6′) and (6″) together are equivalent to (3), if $q^2 \neq -1$.

A similar direct calculation shows that

$$(\xi', \eta') \text{ is a point of } A_q^{0|2} \Leftrightarrow (6'')$$

thus proving (i) $\Leftrightarrow$ (ii). ■

5. Proof of theorem 3. We can now prove the multiplicativity property in a quite natural way. Take a generic point (x, y) of $A_q^{2|0}$ in a ring containing (a, b, c, d) and (a', b', c', d'). We can find such a ring and a point, commuting with (a, b, c, d), (a′, b′, c′, d′), e.g. k[a, ..., d′] $\otimes A_q^{2|0}$ will do. Then $\begin{pmatrix} a & b \\ c & d \end{pmatrix}\begin{pmatrix} x \\ y \end{pmatrix}$ is a point of $M_q(2)$ commuting with (a′, b′, c′, d′). It is also generic since it becomes generic after a specialization $\begin{pmatrix} a & b \\ c & d \end{pmatrix} \to \begin{pmatrix} 1 & 0 \\ 0 & 1 \end{pmatrix}$. Therefore $\begin{pmatrix} a' & b' \\ c' & d' \end{pmatrix}\begin{pmatrix} a & b \\ c & d \end{pmatrix}\begin{pmatrix} x \\ y \end{pmatrix}$ is a point of $M_q(2)$. Hence $\begin{pmatrix} a' & b' \\ c' & d' \end{pmatrix}\begin{pmatrix} a & b \\ c & d \end{pmatrix}$ verifies (6′). Similarly one proves (6″), using $A_q^{0|2}$. This reasoning is valid only for $q^2 \neq -1$, but the result is true also for $q^2 = -1$ in view of the "principle of algebraic identities".

We also get a natural definition of the quantum determinant which immediately proves its multiplicativity: in the notation of Theorem 4,

$$\xi'\eta' = (a\xi + b\eta)(c\xi + d\eta) = \mathrm{DET}_q\begin{pmatrix} a & b \\ c & d \end{pmatrix}\xi\eta. \quad \blacksquare$$

6. Quantum groups GL(2) and SL(2) and their representations. We can now define noncommutative group spaces, imitating the classical procedure of inverting DET_q or putting it equal to 1. (For a justification and a more sophiscated treatment, see §2 and §7.)

$$GL_q(2) : M_q(2)[t] / ([t, a], [t, b], [t, c], [t, d]; t\mathrm{DET}_q - 1),$$

$$SL_q(2) : M_q(2) / (\mathrm{DET}_q - 1).$$

Theorem 4 describes their representations in quantum spaces $A_q^{2|0}$ and $A_q^{0|2}$. There are two ways of looking at these representations. One can consider, say, the whole of $A_q^{2|0}$ as the noncommutative space upon which $GL_q(2)$ acts as an analogue of the fundamental two-dimensional representation. Alternatively, one can consider just its linear part $(A_q^{2|0})_1 = kx \oplus ky$ as the analogue. Then $(A_q^{2|0})_d$ will correspond to the "quantum d–th symmetric power" of the fundamental representation and A to a kind of bosonic Fock space. Actually, these two viewpoints can and should be reconciled in the notion that the quantum d–th symmetric power of the fundamental representation has as its noncommutative function ring the "Veronese subring" $\bigoplus_{i\geq 0} (A_q^{2|0})_{di}$.

7. Further developments. The rest of this paper is an extension of this example in various directions. We shall show that one can replace $A_q^{2|0}$ by an arbitrary quadratic algebra and still define a reasonable quantum semigroup space acting upon it. In order to turn it into a quantum group, we have to work a bit harder than in our example (or in the commutative case) since in general there will be no determinant with the necessary properties.

The whole initial setting can be generalized so as to include, e.g., quantum deformations of linear supergroups. Actually, the correct way to do this is to develop an axiomatic theory of exact tensor categories. In this way, we get a better understanding of the Yang–Baxter (or triangle) equations and their role in the construction of quantum groups.

Finally, we present an initiation to the homological study of quantum spaces and groups.

2. BIALGEBRAS AND HOPF ALGEBRAS

1. Bialgebras. Let H be a k–module. Recall that a bialgebra structure on H is defined by four morphisms

$$H \otimes H \xrightarrow{m} H \xrightarrow{\Delta} H \otimes H,$$

$$k \xrightarrow{\eta} H \xrightarrow{\varepsilon} k,$$

satisfying the following axioms, written as commutative diagrams:

algebras

associativity:

$$\begin{array}{ccccc} & & H\otimes H & & \\ m\otimes id & \nearrow & & \searrow m & \\ H\otimes H\otimes H & & & & H \\ id\otimes m & \searrow & & \nearrow m & \\ & & H\otimes H & & \end{array}$$

unit:

$$\begin{array}{ccc} & H\otimes H & \\ id\otimes\eta,\ \eta\otimes id \nearrow & & \searrow m \\ H = H\otimes k = k\otimes H & \xrightarrow[id]{} & H \end{array}$$

coalgebras

coassociativity:

$$\begin{array}{ccccc} & & H\otimes H & & \\ \Delta & \nearrow & & \searrow & \Delta\otimes id \\ H & & & & H\otimes H\otimes H \\ \Delta & \searrow & & \nearrow & id\otimes\Delta \\ & & H\otimes H & & \end{array}$$

counit:

$$\begin{array}{ccc} & H\otimes H & \\ \Delta \nearrow & & \searrow \varepsilon\otimes id,\ id\otimes\varepsilon \\ H & \xrightarrow[id]{} & H = k\otimes H = H\otimes k \end{array}$$

connecting axiom:

$$\begin{array}{ccccc} & & H & & \\ & m \nearrow & & \searrow \Delta & \\ H\otimes H & & & & H\otimes H \\ \Delta\otimes\Delta \downarrow & & & & \uparrow m\otimes m \\ H\otimes H\otimes H\otimes H & & \xrightarrow{S_{(23)}} & & H\otimes H\otimes H\otimes H \end{array}$$

Here S_σ means the canonical morphism corresponding to a substitution σ. The connecting axiom means that Δ is an algebra morphism or, equivalently, that m is a coalgebra morphism. Its diagrammatic form makes evident its self-dual nature.

2. Antipode. An antipode of a bialgebra (H, m, Δ) is a linear map $i : H \to H$ such that the following diagram is commutative:

$$\begin{array}{ccccccc} & & H\otimes H & \xrightarrow{i\otimes id} & H\otimes H & & \\ & \Delta\nearrow & & & & \searrow m & \\ & H & \xrightarrow{\varepsilon} & k & \xrightarrow{\eta} & H & \\ & \Delta\searrow & & & & \nearrow m & \\ & & H\otimes H & \xrightarrow[id\otimes i]{} & H\otimes H & & \end{array}$$

3. Some elementary constructions. Let $m^{op} = m \circ S_{(12)}$, $\Delta^{op} = S_{(12)} \circ \Delta$. If (H, m, Δ) is a bialgebra, then (H, m^{op}, Δ), (H, m, Δ^{op}) are also bialgebras. If i is a bijective antipode for (H, m, Δ), then i^{-1} is one for (H, m^{op}, Δ) and (H, m, Δ^{op}), hence i is one for (H, m^{op}, Δ^{op}).

4. Theorem. If an antipode i exists, it is unique and reverses multiplication and comultiplication, i.e. it defines a bialgebra morphism (ε, η are not changed):

$$i : (H, m, \Delta) \to (H, m^{op}, \Delta^{op}).$$

In other words, we have commutative diagrams

$$\begin{array}{ccc} & H & \\ m\nearrow & & \searrow i \\ H\otimes H & & H \\ S_{(12)}\downarrow & & \uparrow m \\ H\otimes H & \xrightarrow{i\otimes i} & H\otimes H \end{array} \qquad \begin{array}{ccc} & H & \\ i\nearrow & & \searrow \Delta \\ H & & H\otimes H \\ \Delta\downarrow & & \uparrow S_{(12)} \\ H\otimes H & \xrightarrow{i\otimes i} & H\otimes H \end{array}$$

For a proof, cf. Abe [A].

5. Bialgebras and quantum groups. A ring of polynomial functions on an affine algebraic group G is a bialgebra with an antipode, the comultiplication being induced by the group law $G \times G \to G$ and the antipode by the inversion map $G \to G : x \to x^{-1}$. This bialgebra is commutative. Dropping this condition of commutativity, we get the general notion of a Hopf algebra, which is a formalization of the (so far) intuitive notion of a quantum group.

The point–functor $A \to \mathrm{Hom}_{k\text{-}Alg}(H, A)$ of a Hopf algebra (H, m, Δ, i) verifies the same properties as that of $GL_q(2)$ (cf. Theorem 3, §1):

a) Let $f, g : H \to A$ be two commuting points (i.e., $[f(h), g(h')] = 0$ for all $h, h' \in H$). Then one can define their product as a composite map:

$$fg : H \xrightarrow{\Delta} H \otimes H \xrightarrow{f \otimes g} A \otimes A \xrightarrow{m_A} A.$$

(One needs commutativity to prove that this is an algebra morphism.)

b) Let $f : H \to A$ be a point. Then, if i is bijective,

$$f \circ i : H \xrightarrow{i} \dot{H} \xrightarrow{f} A$$

is a point of $(H, m^{op}, \Delta^{op}, i)$. (Reversing multiplication and comultiplication for $GL_q(2)$, we get $GL_{q^{-1}}(2)$.)

Note that in a general bialgebra (i.e. "quantum semigroup") i may not exist (as in $M_q(2)$); if it exists, it may not be bijective. If it is bijective, it may happen that $i^2 \neq \mathrm{id}$.

6. Multiplicative matrices. Let (H, Δ) be a coalgebra with a counit ε, $Y \in M(n, H)$ a matrix with elements in H. We shall call $Y = (y_i^j)$ multiplicative if

$$\Delta(Y) = Y \otimes Y \quad , \quad \varepsilon(Y) = E. \tag{1}$$

This is not a standard notation for the tensor product of matrices. It simply means that

$$\Delta(y_i^j) = \sum_k y_i^k \otimes y_k^j \quad , \quad \varepsilon(y_i^j) = \delta_i^j.$$

Examples: $\begin{pmatrix} a & b \\ c & d \end{pmatrix}$ in $M_q(2)$, $\begin{pmatrix} a & b & 0 \\ c & d & 0 \\ 0 & 0 & t \end{pmatrix}$ in $GL_q(2)$. If a quantum group H admits a multiplicative matrix Y whose entries generate H as a ring, H can be called a "quantum matrix group" (cf. Woronowicz [W1], [W2], who also uses a special sign instead of our fake tensor product). We shall give below a representation–theoretic interpretation of multiplicative matrices, but first we state some of their properties, which will help us in §7 to construct antipodal maps.

7. Proposition.

a) $\Delta(Y) = Y \otimes Y \Leftrightarrow \Delta^{op}(Y^t) = Y^t \otimes Y^t$.

b) Assume that Y is a multiplicative matrix in a Hopf algebra with antipode i. Define $Y_k = i^k(Y)$. Then

$$\begin{aligned} Y_k Y_{k+1} &= Y_{k+1} Y_k = I \text{ for } k \equiv 0 \pmod 2, \\ Y_k^t Y_{k+1}^t &= Y_{k+1}^t Y_k^t = I \text{ for } k \equiv 1 \pmod 2, \end{aligned} \tag{2}$$

and

$$\Delta(Y_k) = \begin{cases} Y_k \otimes Y_k & \text{for } k \equiv 0 \pmod 2 \\ (Y_k^t \otimes Y_k^t)^t & \text{for } k \equiv 1 \pmod 2. \end{cases} \tag{3}$$

c) If f, g are commuting A-points of H, as in no. 5, then

$$(fg)(Y) = f(Y)g(Y).$$

Proof.

a) Let $\Delta(Y) = Y \otimes Y$. Then

$$(\Delta^{op}(Y^t))^{t}{}_i^k = \Delta^{op}(Y)_k^i = S_{(12)} \circ \Delta(Y)_k^i =$$

$$= S_{(12)}(\sum_j y_k^j \otimes y_j^i) = \sum_j y_j^i \otimes y_k^j;$$

$$(Y^t \otimes Y^t)^{t}{}_i^k = \sum_j (Y^t)^{t}{}_i^j \otimes (Y^t)^{t}{}_j^k = \sum_j y_j^i \otimes y_k^j.$$

b) Applying the antipode axiom (no. 2) to Y, we get

$$i(Y)Y = Y\,i(Y) = I.$$

Since i reverses multiplication, we have $i(AB) = [i(B^t)i(A^t)]^t$ for two matrices A, B in H. From this, we obtain (2) by induction. Finally, again by induction

$$\Delta(Y_{k+1}) = \Delta \circ i(Y_k) =$$

$$= S'_{(12)}(i \otimes i)(\Delta(Y_k)) = \begin{cases} S_{(12)}(Y_{k+1} \otimes Y_{k+1}) & \text{for } k \equiv 0 \pmod 2 \\ S_{(12)}[(Y_{k+1}^t \otimes Y_{k+1}^t)]^t & \text{for } k \equiv 1 \pmod 2. \end{cases}$$

This establishes (3) since the computation at the beginning of our proof shows that $S_{(12)}(Y_k \otimes Y_k) = (Y_k^t \otimes Y_k^t)^t$.

c) By definition, and (1)

$$(fg)(Y) = m_A \circ (f \otimes g) \circ \Delta(Y) = f(Y)g(Y). \quad \blacksquare$$

8. Comodules. A left comodule over a coalgebra (H, Δ, ε) is a linear space M together with a morphism $\delta : M \to H \otimes M$ such that the following two diagrams are commutative:

coassociativity:

$$\begin{array}{ccc} & H \otimes M & \\ \delta \nearrow & & \searrow \Delta \otimes id \\ M & & H \otimes H \otimes M \\ \delta \searrow & & \nearrow id \otimes \delta \\ & H \otimes M & \end{array}$$

counit:

$$\begin{array}{ccc} & H \otimes M & \\ \delta \nearrow & & \searrow \varepsilon \otimes id \\ M & \underset{id}{\rightarrow} & k \otimes M = M \end{array}$$

Similarly one defines a right comodule via $\delta : M \to M \otimes H$.

Let (M, δ) be an (H, Δ, ε)–left–comodule. Then $(M, \delta^{op} = S_{(12)} \circ \delta)$ is an $(H, \Delta^{op}, \varepsilon)$–right–comodule, and vice versa. A morphism of comodules is defined in the natural way.

Let (k^n, δ) be a finite–dimensional left H–comodule. Define a matrix Y by:

$$\delta(e_i) = \sum_{j=1}^{n} y_i^{\ j} \otimes e_j,$$

where e_j is a canonical basis of k^n.

9. Proposition.

a) This construction establishes a bijection between all structures of a left comodule on k^n and multiplicative matrices in $M(n, H)$.

b) A linear map $f : (k^n, \delta) \to (k^m, \delta')$ is a morphism of comodules given by multiplicative matrices Y, Y', iff

$$FY' = YF,$$

where $F = (f_i^j)$, $f(e_i) = \sum_j f_i^j e_j'$.

Proof. Applying to (k^n, δ) the coassociativity axiom, we get $\Delta(Y) = Y \otimes Y$. Similarly, applying the counit axiom, we get $\varepsilon(Y) = Y$. Clearly, these conditions also imply the axioms. The last statement is straightforward. ■

10. Representations. In general, there are at least three notions of a representation of a quantum (semi) group H.

a) A (left or right) H–comodule.

This corresponds to the common notion of group representation if one looks at H as the function ring of our group. We adopt this viewpoint further on.

b) A (left or right) H–module.

One rarely considers this notion in the classical situation since (H, m) is then commutative. However, the classical representations are modules over the universal enveloping algebras dual to function rings. Thus, morally, H–modules are representations of a "dual quantum group".

c) A morphism of Hopf algebras $H' \to H$ (or, dually $H \to H'$).

This viewpoint corresponds to the classical notion of, say, a unitary representation considered as a morphism $G \to U(n)$.

Below, we shall construct quantum endomorphism semigroups (resp. automorphism groups) of general quadratic algebras which will provide us with representations in both senses a) and c), but not b).

11. Generators and relations. Algebras are often defined by generators and relations, i.e. as, say, $T(V)/I$ where V is a linear space, $T(V)$ its tensor algebra and I the ideal of relations.

In general, a linear subspace $I \subset H$ is an ideal in (H, m) if $m(I \otimes H + H \otimes I) \subset I$. Dually, it is a coideal in the coalgebra (H, Δ) if $\Delta(I) \subset I \otimes H + H \otimes I$. A subspace I in a bialgebra which is both an ideal and a coideal induces a bialgebra structure on H/I. An antipode i_H survives in H/I if $i_H(I) \subset I$.

3. QUADRATIC ALGEBRAS AS QUANTUM LINEAR SPACES

1. Notation. A quadratic algebra is an associative graded algebra $A = \bigoplus_{i=0}^{\infty} A_i$ with the following properties: $A_0 = k$ (ground field); A is generated by A_1; the ideal of relations among elements of A_1 is generated by the subspace of all quadratic relations $R(A) \subset A_1^{\otimes 2}$. It is convenient to write $A \leftrightarrow \{A_1, R(A)\}$. We assume $\dim A_1 < \infty$.

Quadratic algebras form a category QA: morphisms $f : A \to B$ are in bijection with linear maps $f_1 : A_1 \to B_1$ such that $(f_1 \otimes f_1)(R(A)) \subset R(B)$. Therefore, we have a forgetful functor $QA \to k\text{-mod} : A \to A_1$.

In the next sections, we shall show that an arbitrary quadratic algebra can play the role of a "quantum plane" of §1.

2. Operations on quadratic algebras. Define for $A, B \in Ob/QA$:

$$\tilde{A} \leftrightarrow \{A_1, \{0\}\}, \tag{1}$$

$$A^{op} \leftrightarrow \{A_1, S_{(12)}(R(A))\}, \tag{2}$$

$$A^{!} \leftrightarrow \{A_1^{*}, R(A)^{\perp}\}, \tag{3}$$

$$A \circ B \leftrightarrow \{A_1 \otimes B_1, S_{(23)}(R(A) \otimes B_1^{\otimes 2} + A_1^{\otimes 2} \otimes R(B))\}, \tag{4}$$

$$A \bullet B \leftrightarrow \{A_1 \otimes B_1, S_{(23)}(R(A) \otimes R(B))\}, \tag{5}$$

$$A \otimes B \leftrightarrow \{A_1 \oplus B_1, R(A) \oplus [A_1, B_1] \oplus R(B)\}, \tag{6}$$

$$A \underline{\otimes} B \leftrightarrow \{A_1 \oplus B_1, R(A) \oplus [A_1, B_1]_+ \oplus R(B)\}. \tag{7}$$

Clearly, $\tilde{A}$ is the tensor algebra of A_1. The natural map $\tilde{A} \to A$, identical on A_1, identifies A with $\tilde{A}/R_A$, where $R_A = \bigoplus_{i=0}^{\infty} R_i(A)$, $R_0(A) = \{0\}$, $R_1(A) = \{0\}$, $R_2(A) = R(A)$,

$$R_n(A) = \sum_{i=0}^{n-2} A_1^{\otimes i} \otimes R(A) \otimes A_1^{\otimes n-i-2}. \qquad (8)$$

In (3), we define $A_1^* = \mathrm{Hom}_k(A_1, k)$, identity $(V \otimes W)^*$ with $V^* \otimes W^*$ by $(f \otimes g)(a \otimes b) = f(a)g(b)$ and set $R(A)^{\perp} = \{q \in A_1^* \otimes A_1^* \mid q(r) = 0\}$ for all $r \in R(A)$.

In (2), (4), (5), S_σ means a map

$$S_\sigma(a_1 \otimes ... \otimes a_n) = a_{\sigma^{-1}(1)} \otimes ... \otimes a_{\sigma^{-1}(n)}.$$

In (6), the commutator $[A_1, B_1]$ (resp. $[A_1, B_1]_+$) means the subspace in $A_1 \otimes B_1 \oplus B_1 \otimes A_1$ generated by $a \otimes b - b \otimes a$ (resp. $a \otimes b + b \otimes a$).

Examples from §1:

$$(A_q^{2|0})^! \cong A_q^{0|2};$$

$$M_q(2) \cong A_q^{0|2} \bullet A_q^{2|0} \quad (6'').$$

One can look at these operations as natural lifts of standard functors from k–mod to QA. Some lifts are double, being interchanged by the duality map !:

QA:	$\sim$	op	!	$\circ \overset{!}{\leftrightarrow} \bullet$	$\otimes \overset{!}{\leftrightarrow} \underline{\otimes}$
	$\downarrow$	$\downarrow$	$\downarrow$	$\searrow\swarrow$	$\searrow\swarrow$
k–mod:	id	id	$*$	$\otimes$	$\oplus$

For the sake of completeness, we shall list below the main interrelations between these functors.

3. Properties of $\sim$. It is a covariant functor $QA \to QA$; the canonical map $\tilde{A} \to A$ is a functor morphism: $\sim \,\to \mathrm{id}$. Moreover,

$$(\tilde{A})^{op} = (A^{op})^{\sim} = \tilde{A}; \ (A^{!})^{\sim} = \tilde{A}^{*} := \bigoplus_i (\tilde{A}_i)^{*}; \ (\tilde{A})^{!} = k \oplus A_1^{*}.$$

Finally,

$$(A \circ B)^{\sim} = (A \bullet B)^{\sim} = A^{\sim} \circ B^{\sim} = A^{\sim} \bullet B^{\sim} = \text{tensor algebra of } A_1 \otimes B_1,$$

$$(A \otimes B)^{\sim} = (A \underline{\otimes} B)^{\sim} = \text{tensor algebra of } A_1 \oplus B_1.$$

4. Properties of op. The map $A \to A^{op}$, $f \to f^{op}$, where $(f^{op})_1 = f_1$, is a covariant involution of QA. Denote by A° the ring, coinciding with A as a linear space, with reverse multiplication: $f * g$ (in A°) $= gf$ (in A). Then the map $\tau : \tilde{A} \to \tilde{A}$, $\tau(a_1 ... a_n) = a_n ... a_1$ induces an isomorphism $A^{op} \xrightarrow{\sim} A^{\circ}$. Furthermore, there are functorial identifications

$$(A^{op})^{!} = (A^{!})^{op}, \quad (A * B)^{op} = A^{op} * B^{op},$$

where $*$ is one of the products (4)–(7).

5. Properties of !. The dualization functor $A \to A^{!}$, $f \to f^{!}$, where $f_1^{!} = f_1^{*} : B_1^{*} \to A_1^{*}$, is a contravariant quasi-involution of QA: !! is equivalent to id. There are natural identifications

$$(A \circ B)^{!} = A^{!} \bullet B^{!}; \qquad (A \bullet B)^{!} = A^{!} \circ B^{!};$$

$$(A \otimes B)^{!} = A^{!} \underline{\otimes} B^{!}; \qquad (A \underline{\otimes} B)^{!} = A \otimes B.$$

6. Properties of products. The multiplications (4)–(7) with the constraints of associativity and commutativity, which are evident on 1–components, define on QA four different structures of tensor categories (cf. Deligne-Milne [DM] and §12 for a more detailed discussion). All these tensor categories have unit objects:

$$K = k[\varepsilon], \ \varepsilon^2 = 0, \text{ for } \circ; \quad L = k[t] = K^{!} \text{ for } \bullet; \quad k \text{ for } \otimes \text{ and } \underline{\otimes}.$$

However, in the category of "quantum linear spaces" QA^{op}, $\bullet$ and $\circ$ are similar to the tensor product while $\otimes$ and $\underline{\otimes}$ correspond to the direct sum.

We have also functorial homomorphisms:

$$A \bullet B \overset{\alpha}{\to} A \circ B \overset{\beta}{\to} A \otimes B,$$

$$\alpha_1 = \mathrm{id} : A_1 \otimes B_1 \to A_1 \otimes B_1;$$

$$\beta(a \otimes b) = a \otimes b \text{ for } a \in A_1, b \in B_1.$$

Note that β is not a morphism in QA since it doubles the degree (if one makes a common convention that $(A \otimes B)_n = \oplus A_i \otimes B_{n-i}$).

7. Lemma. β induces an isomorphism of rings

$$A \circ B \overset{\sim}{\to} \sum_{n=0}^{\infty} A_n \otimes B_n \subset A \otimes B.$$

Proof. We must check that, in the notation of no. 2,

$$R_n(A \circ B) = S_\sigma(R_{2n}(A \otimes B) \cap A_1^{\otimes n} \otimes B_1^{\otimes n}).$$

where σ is the substitution transforming $A_1^{\otimes n} \otimes B_1^{\otimes n}$ into $(A_1 \otimes B_1)^{\otimes n}$, and conserving the relative order of A– and B–factors. But from (8), one sees that

$$R_n(A \circ B) = \sum_{i=1}^{n} (A_1 \otimes B_1)^{\otimes n} \otimes S_{(23)}[R(A) \otimes B_1^{\otimes 2} + A_1^{\otimes 2} \otimes R(B)] \otimes (A_1 \otimes B_1)^{\otimes(n-2-i)}$$

$$R_{2n}(A \otimes B) \cap A_1^{\otimes n} \otimes B_1^{\otimes n} =$$

$$= \sum_{i=1}^{n} (A_1 \oplus B_1)^{\otimes 2i} \otimes [R(A) \oplus [A_1, B_1] \oplus R(B)] \otimes (A_1 \oplus B_1)^{\otimes 2n-2i-2} \cap A_1^{\otimes} \otimes B_1^{\otimes n}$$

and a reshuffling of factors proves the lemma. ■

8. Quantum symmetric power. For $A \in \mathrm{Ob}QA$ define

$$A^{(d)} = \bigoplus_{i=0}^{\infty} A_{nd}.$$

By analogy with the commutative polynomial case, we shall call $A^{(d)}$ (the function ring of) the d–th symmetric power of (the quantum space defined by) A.

9. Proposition. $A^{(d)}$ is a quadratic algebra.

This follows easily from (8). ■

In fact, Backelin and Fröberg [BF] proved that if A is a graded algebra, generated by finite–dimensional A_1, and if the ideal of relations is generated by its components of degree $\leq r$, then the same is true for $A^{(d)}$, with $[2 + (r-2)/d]$ instead of r. Hence any finitely related and finitely generated graded algebra gives rise to a quadratic algebra under an operation $A \to A^{(d)}$. In the commutative case, this operation does not change Proj A. Therefore quadratic algebras essentially exhaust the range of projective algebraic geometry.

10. Quantum exterior power. Since $S^{\bullet}(V)^{!} = \Lambda^{\bullet}(V^{*})$, we can call $A^{!(d)}$ the d–th exterior power of the dual space of A. However, it is better to reinterprete ! using the lessons of superalgebras: it is a functor which combines dualization and parity change. Without imposing some additional structures, we cannot define these two constructions separately.

11. Résumé. Looking at QA^{op} as the category of "quantum linear spaces", we shall utilize the following analogies:

$\circ$: tensor product of quantum spaces;

$\otimes$: direct sum of quantum spaces;

! : dualization + parity change;

(d) : d–th symmetric power.

4. QUANTUM MATRIX SPACES I.

CATEGORICAL VIEWPOINT

1. Motivation. Let U, V be finite dimensional linear spaces. Then we have the natural maps

$$\mathrm{Hom}(U, V) \otimes U \to V : f \otimes u \to f(u); \qquad (1)$$

$$\mathrm{Hom}(V, T) \otimes \mathrm{Hom}(U, V) \to \mathrm{Hom}(U, T) : f \otimes g \to f \bullet g, \qquad (2)$$

with well–known universality properties.

Passing to the polynomial function rings $A(V) = S(V^*)$ etc, we get dual maps

$$A(V) \to A(\mathrm{Hom}(U, V)) \circ A(U); \qquad (3)$$

$$A(\mathrm{Hom}(U, T)) \to A(\mathrm{Hom}(V, T)) \circ A(V, T).. \qquad (4)$$

We shall show that in QA, we also have natural maps (3) and (4), if we define

$$A(\mathrm{Hom}(U, V)) := A(U)^! \bullet A(V). \qquad (5)$$

Quadratic algebras $A^! \bullet B$ thus will play a role of quantum matrix spaces (see, however, a warning at the end of this section).

Here we shall clarify the categorical content of this construction, while in the next sections, we investigate quantum semigroups $A^! \bullet A$ and the problem of their extension to quantum groups using coordinates.

2. Theorem. There is a functorial isomorphism

$$\mathrm{Hom}(A \bullet B, C) = \mathrm{Hom}\,(A, B^! \circ C)$$

identifying a map $f : A_1 \otimes B_1 \to C_1$ with a map $g : A_1 \to B_1^* \otimes C_1$, if $\langle g(a) \mid b \rangle = f(a \otimes b)$ for all $a \in A_1$, $b \in B_1$ (the l.h.s. denotes contraction with respect to B_1).

Corollary. (QA, •) is a (non–additive) tensor category with internal $\underline{Hom}$ and unit object $K = k[\varepsilon]$, $\varepsilon^2 = 0$. ■

Proof. We must check that if f, g are related as in the statement, the following conditions are equivalent:

$$(f\otimes f)S_{(23)}(R(A)\otimes R(B)) \subset R(C),$$

$$(g\otimes g)R(A) \subset S_{(23)}(R(B)^{\perp}\otimes C_1^{\otimes 2} + B_1^{*\otimes 2}\otimes R(C)).$$

But they are respectively equivalent to:

$\langle R(C)^{\perp} \mid (f\otimes f)S_{(23)}(R(A)\otimes R(B))\rangle = 0$ (contraction w.r.t. $C_1\otimes C_1$),

$\langle R(B)\otimes R(C)^{\perp} \mid (g\otimes g)R(A)\rangle = 0$ (contraction w.r.t. $C_1\otimes C_1\otimes C_1^*\otimes C_1^*$).

Each of these last orthogonality relations means that if we start with an element of R(A), apply $g\otimes g$ and then contract consecutively with arbitrary elements of R(B) and $R(C)^{\perp}$, we get zero. ■

3. Internal Hom. Following the general formalism of tensor categories [DM], we define $\underline{Hom}(B, C) = B^! \circ C$. In particular, $B^! = \underline{Hom}(B, K^!)$. Therefore, it is n̲o̲t̲ the standard dualization in (QA, •) which would be $\check{B} = \underline{Hom}(B, K) = k\otimes B_1^*$, hardly a very interesting object.

By the general properties of $\underline{Hom}$, the following internal product maps are defined:

$$\beta : \underline{Hom}(B, C)\bullet B \to C \tag{6}$$

$$\mu : \underline{Hom}(C, D)\bullet \underline{Hom}(B, C) \to \underline{Hom}(B, D), \tag{7}$$

with obvious associativity properties.

The map (6) verifies the following universality property: for any morphism $f : A \bullet B \to C$ in QA, there exists a unique morphism $g : A \to \underline{Hom}(B, C)$ making the following diagram commutative:

$$\begin{array}{ccc} \underline{Hom}(B, C) \bullet B & \overset{\beta}{\to} & C \\ g \bullet id \uparrow & \nearrow f & \\ A \bullet B & & \end{array}$$

In fact, this map is just the identification defined in Theorem 2. One can then construct (7) iterating (6).

4. Internal hom. Comparing (6) and (7) with (3) and (4), we see however that $\underline{Hom}(B, C)$ is not the quantum matrix space we are looking for. In fact, it is dual to it with respect to !.

Define

$$\underline{hom}(B, C) = \underline{Hom}(B^!, C^!)^! = B^! \bullet C.$$

Write (6), (7) for $B^!$, $C^!$, $D^!$ and apply !. We get morphisms

$$\delta : C \to \underline{hom}(B, C) \circ B, \tag{8}$$

$$\Delta : \underline{hom}(B, D) \to \underline{hom}(C, D) \circ \underline{hom}(B, C) \tag{9}$$

defined by a universality property dual to the one in no 3.

5. Theorem.

a) For any morphism $g : C \to A \circ B$ in QA there exists a unique morphism $f : \underline{hom}(B, C) \to A$ such that $g = (f \circ id_B)\delta$.

b) Let $B = C = D$ in (8) and (9). Then $E = \underline{end}(B) = \underline{hom}(B, B) = B^! \bullet B$ becomes a bialgebra with respect to the following data:

m_E = multiplication;

Δ_E = composition $E \overset{(9)}{\to} E \circ E \to E \otimes E$;

η : $k = E_0 \overset{id}{\tilde{\to}} k$;

ε : $E_1 = B_1^* \otimes B_1 \to k$ the standard pairing

c) The map (8) $\delta : B \to E \circ B$ induces on B (and on all homogeneous components B_i of B) a structure of left E–comodule. ■

Part a) of this theorem is again a restatement of Theorem 3.

We shall prove parts b) and c) in the next section using coordinate language, and we shall return to the categorical picture in §12.

6. Warning. If we apply our constructions to the usual linear spaces, i.e., polynomial rings, we do not obtain matrices with commuting entries !.

In fact, end $k[x_1,...,x_n]$ is a very noncommutative ring: it is defined only by half of the necessary commutation relations. In order to get the other half, we may impose the same commutation relations upon the transposed matrix, as in the proof of Theorem 4, §1. See the next section for details.

5. QUANTUM MATRIX SPACES II.

COORDINATE APPROACH

1. A quantum space. Consider a general quadratic algebra

$$A = k\langle \tilde{x}_1, \ldots, \tilde{x}_n \rangle / (r_\alpha)$$

where $k\langle \tilde{x} \rangle$ means a free associative algebra generated by $\tilde{x}_i$, and

$$r_\alpha = r_\alpha(\tilde{x}) = \sum_{ij} c_\alpha^{ij} \tilde{x}_i \tilde{x}_j \ , \ \alpha = 1, \ldots, m \tag{1}$$

are linearly independent elements of $k\langle \tilde{x}_1, \ldots, \tilde{x}_n \rangle_2$. We define $R = (r_\alpha)$ and $x_i = \tilde{x}_i \bmod R$; R may also mean relations in any algebra to appear later.

2. The dual space. This is defined by the algebra

$$A^! = k\langle \tilde{x}^1, \ldots, \tilde{x}^n \rangle / (r^\beta)$$

$$r^\beta = \sum_{k\ell} c_{k\ell}^\beta \tilde{x}^k \tilde{x}^\ell, \beta = 1, \ldots, n^2 - m \tag{2}$$

where $\langle \tilde{x}^i | \tilde{x}_j \rangle = \delta_j^r$ and $\oplus k r^\beta = (\oplus k r_\alpha)^\perp$, i.e.,

$$\langle r^\beta | r_\alpha \rangle = \sum_{ij} c_{ij}^\beta c_\alpha^{ij} = 0 \tag{3}$$

We again let $x^i = \tilde{x}^i \bmod R$.

3. The matrix space end (A). We have

$$\underline{\text{end}}\,(A) = A^! \bullet A = k\langle \tilde{z}_i^k \rangle / (r_\alpha^\beta);$$

$$\tilde{z}_i^k = \tilde{x}^k \otimes \tilde{x}_i \quad ; \quad z_i^k = \tilde{z}_i^k \bmod R;$$

$$r_\alpha^\beta = S_{(23)}(r^\beta \otimes r_\alpha) = \sum_{ijk\ell} c_\alpha^{ij} c_{k\ell}^\beta \tilde{z}_i^k \tilde{z}_j^\ell \tag{4}$$

4. Action. This is an algebra map (a morphism in QA)

$$\delta_A : A \to \underline{\text{end}}\,(A) \circ A \subset \underline{\text{end}}\,(A) \otimes A,$$

$$\delta_A(x_i) = \sum_k z_i^k \otimes x_k \qquad (5)$$

5. Lemma. (5) is well-defined. In fact, let

$\tilde{\delta}: k\langle \tilde{x}_1, \ldots, \tilde{x}_n \rangle \to k\langle \tilde{z}_i^k \rangle \otimes k\langle \tilde{x}_j \rangle$ be defined by the same formula (5). Then, for some $s_\beta(\tilde{x}) \in k\langle \tilde{x} \rangle$, $s_\alpha^\beta(\tilde{z}) \in k\langle \tilde{z} \rangle$ we have

$$\tilde{\delta}(r_\alpha) = \sum_\beta r_\alpha^\beta(\tilde{z}) \otimes s_\beta(\tilde{x}) + \sum_{\alpha'} s_\alpha^{\alpha'}(\tilde{z}) \otimes r_{\alpha'}(\tilde{x}) \qquad (6)$$

Proof. Choose $n^2 - m$ quadratic forms $s_\beta = s_\beta(\tilde{x}) \in k\langle \tilde{x} \rangle$ in such a way that $\langle r^\beta | s_\gamma \rangle = \delta_\gamma^\beta$. Then $\{r_\alpha, s_\beta\}$ form a basis of $k\langle \tilde{x} \rangle_2$. We affirm that

$$\tilde{x}_k \tilde{x}_\ell = \sum_\beta c_{k\ell}^\beta s_\beta + \sum_\alpha e_{k\ell}^\alpha r_\alpha$$

for $c_{k\ell}^\beta$ from (2) and some constants $e_{k\ell}^\alpha$. This is checked by taking scalar products with r^β. Now,

$$\tilde{\delta}(r_\alpha) = \sum_{ij} c_\alpha^{ij} \Big(\sum_k \tilde{z}_i^k \otimes \tilde{x}_k\Big)\Big(\sum_\ell \tilde{z}_j^\ell \otimes \tilde{x}_\ell\Big) =$$

$$= \sum_{ijk\ell} c_\alpha^{ij} \tilde{z}_i^k \tilde{z}_j^\ell \otimes \Big(\sum_\beta c_{k\ell}^\beta s_\beta + \sum_{\alpha'} e_{k\ell}^{\alpha'} r_{\alpha'}\Big) =$$

$$= \sum_{\beta ijk\ell} c_\alpha^{ij} c_{k\ell}^\beta \tilde{z}_i^k \tilde{z}_j^\ell \otimes s_\beta(\tilde{x}) + \sum_{\alpha' ijk\ell} c_\alpha^{ij} e_{k\ell}^{\alpha'} \tilde{z}_i^k \tilde{z}_j^\ell \otimes r_{\alpha'}(\tilde{x})$$

which proves (6) in view of (4). ∎

6. Lemma on universality. For any k-algebra B and morphism $\delta: A \to B \otimes A$ with $\delta(A_1) \subset B \otimes A_1$ there exists a unique morphism $\gamma: \mathrm{end}(A) \to B$ such that the diagram

$$\begin{array}{ccc} A & \xrightarrow{\delta} & B \otimes A \\ & \delta_A \searrow & \uparrow \gamma \otimes \mathrm{id} \\ & & \mathrm{end}(A) \otimes A \end{array}$$

commutes. If, moreover, δ is a morphism of quadratic algebras $\delta: A \to B \circ A \to B \otimes A$, then γ is also a morphism in QA.

Proof. Let $v_i^k \in B$ be defined by $\delta(x_i) = \sum v_i^k \otimes x_k$. Putting x_i, v_i^k into the universal formula (6) instead of $\tilde{x}_i, \tilde{z}_i^k$, we get

$$0 = \delta(r_\alpha(x_i)) = \sum_\beta r_\alpha^\beta(v) \otimes s_\beta(x).$$

Hence $r_\alpha^\beta(v) = 0$ since $s_\beta(x)$ are linearly independent in A_2. Therefore we can define $\gamma: \mathrm{end}\,(A) \to B$ by $\gamma(z_i^k) = v_i^k$. The rest is clear. ■

7. The diagonal map. Using universality, we can now define the diagonal map $\Delta_A: E \to E \circ E \to E \otimes E$, $E = \mathrm{end}\,(A)$, by the commutative diagram.

$$\begin{array}{ccccc} A & \xrightarrow{\delta_A} & E \otimes A & \xrightarrow{\mathrm{id} \otimes \delta_A} & E \otimes E \otimes A \\ & \delta_A \searrow & & \nearrow \Delta_A \otimes \mathrm{id} & \\ & & E \otimes A & & \end{array}$$

In this way, the coassociativity axiom for δ_A (§2.8) becomes evident.

Applying morphisms in this diagram to x_i, we get

$$\Delta(z_i^k) = \sum_j z_i^j \otimes z_j^k,$$

or, as we were writing it in §2.6, $\Delta(Z) = Z \otimes Z$, $Z = (z_i^k)$.

8. The end of the proof of theorem 4.5. We have already defined most of the data, describing E and the action of E upon A, and checked most of the axioms. We have not yet mentioned the counit. Define $z_i^k = \delta_i^k$. We have $r_\alpha^\beta(\delta_i^k) = \delta_i^k$ in view of (3) so that the relations in E are not violated. The counit axioms for Δ_A and δ_A are now evident.

For subsequent constructions it is useful to keep in mind an analogue of Lemma 5 for Δ_A: if $\tilde{\Delta}(\tilde{Z}) = \tilde{Z} \otimes \tilde{Z}$, then

$$\tilde{\Delta}(r_\alpha^\beta) = \sum_{\gamma\delta} r_\gamma^\delta(\tilde{z}) \otimes s_{\alpha\delta}^{\beta\gamma}(\tilde{z}) + \sum_{\gamma\delta} t_{\alpha\delta}^{\beta\gamma}(\tilde{z}) \otimes r_\gamma^\delta(\tilde{z}) \qquad (7)$$

for some $s, t \in k\langle \tilde{z}_i^k \rangle$. (This follows from what we have already proved, but can also be proved directly.)

We shall use formulas of this kind in order to check that an ideal in a bialgebra is in fact a coideal and therefore, it can be factored out consistently with the codiagonal map.

We shall now discuss the functorial behaviour of end (A).

9. Dualization. In commutative geometry, end L $\cong$ end L* canonically as linear spaces, but multiplication is reversed. Exactly the same holds in our situation.

10. Theorem. There exists a canonical isomorphism of bialgebras $\tau_A: (\underline{end}\,(A), m_A, \Delta_A) \tilde{\to} (\underline{end}(A^!), m_{A^!}, \Delta^{op}_{A^!})$ coinciding with $S_{12}: A_1^* \otimes A_1 \to A_1 \otimes A_1^*$ on 1-components.

Proof. We shall work it out in coordinates. Put $\check{x}_i = \tilde{x}^i$, $\check{x}^i = \tilde{x}_i$ thus identifying $A^{!!} = A$. Then

$$A^! = k\langle \check{x}_i \rangle / (\check{r}_\beta)$$

$$\check{r}_\beta = \sum_{ij} \check{c}^{ij}_\beta \check{x}_i \check{x}_j \quad , \quad \check{c}^{ij}_\beta = c^\beta_{ij} \quad , \quad \beta = 1, \dots, n^2 - m$$

Similarly,

$$\check{r}^\alpha = \sum_{ij} \check{c}^\alpha_{k\ell} \check{x}^k \check{x}^\ell \quad , \quad \check{c}^\alpha_{k\ell} = c^{k\ell}_\alpha \quad , \quad \alpha = 1, \dots, m$$

Therefore

$$\underline{end}\,(A^!) = A \bullet A^! = k\langle \check{z}^k_i \rangle / (\check{r}^\alpha_\beta),$$

$$\check{r}^\alpha_\beta = S_{(23)}(\check{r}^\alpha \otimes \check{r}_\beta) = \sum \check{c}^{ij}_\beta \check{c}^\alpha_{k\ell} \check{z}^k_i \check{z}^\ell_j. \tag{8}$$

Let $\tilde{Z} = (\tilde{z}^k_i)$, $\check{Z} = (\check{z}^k_i)$. Comparing (8) with (4), one sees that the map $\tilde{\tau}: \tilde{Z} \to \check{Z}^t$ verifies $\tilde{\tau}(r^\beta_\alpha) = \check{r}^\alpha_\beta$ and thererefore defines an algebra isomorphism $\underline{end}\,(A) \tilde{\to} \underline{end}\,(A^!)$.

In other words, $\check{r}$ can be obtained from r by transposing Z.

Finally,

$$\Delta^{op}_{A^!}(\check{Z}^t) = \check{Z}^t \otimes \check{Z}^t$$

in view of Proposition 2.7a), since $\Delta_{A^!}(\check{Z}) = \check{Z} \otimes \check{Z}$. Hence $\tilde{\tau}$ induces the bialgebra isomorphism.

Clearly, $\tilde{\tau}$ coincides with S_{12} on generators. Still, we shall write the coordinate change formulas for future use. If

$$x' = \begin{pmatrix} x'_1 \\ \cdot \\ \cdot \\ \cdot \\ x'_n \end{pmatrix} = U \begin{pmatrix} x_1 \\ \cdot \\ \cdot \\ \cdot \\ x_n \end{pmatrix}$$

in A_1, $U \in GL(n,k)$, then

$$\check{x}' = (U^t)^{-1} x,$$

$$Z' = UZU^{-1}, \ \check{Z}' = (U^t)^{-1} \check{Z} U^t \qquad (9)$$

and one sees again that τ_A is well-defined. ■

11. Opposite bialgebra. In coordinates, we have:

$$A^{op} = k\langle \overset{o}{x}_1, \ldots, \overset{o}{x}_n \rangle / (\overset{o}{r}_\alpha)$$

$$\overset{o}{r}_\alpha = \sum \overset{o}{c}{}^{ij}_\alpha \overset{o}{x}_i \overset{o}{x}_j \quad , \quad \overset{o}{c}{}^{ij}_\alpha = c^{ji}_\alpha .$$

Therefore, in self–evident notation

$$\underline{end}\,(A^{op}) = k\langle \overset{o}{z}{}^k_i \rangle / (\overset{o}{r}{}^\beta_\alpha),$$

$$\overset{o}{r}_\alpha = \sum \overset{o}{c}{}^{ij}_\alpha \overset{o}{x}_i \overset{o}{x}_j, \quad \overset{o}{c}{}^{ij}_\alpha = c^{ji}_\alpha .$$

12. Theorem. The map $\tilde{\rho} : \tilde{Z} \to \overset{o}{Z}$, $\tilde{\rho}(fg) = \tilde{\rho}(g)\tilde{\rho}(f)$ verifies $\tilde{\rho}(r^\beta_\alpha) = \overset{o}{r}{}^\beta_\alpha$ and therefore it induces a canonical bialgebra isomorphism

$$\rho_A : (\underline{end}\,(A), m_A, \Delta_A) \to (\underline{end}\,(A^{op}), m_{A^{op}}, \Delta_{A^{op}}).$$

Proof. Clear. ■

13. Symmetric power. As explained earlier, we may look at A_1 as the basic (co)representation of $\underline{end}(A)$ and at A_d as its symmetric power. On the other hand, it is more suggestive to imagine the whole quantum space A as the fundamental representation and $A^{(d)}$ as its d-th symmetric power. Then we have the action

$$\delta_A|_{A^{(d)}} : A^{(d)} \to end\ (A) \otimes A^{(d)}.$$

Since clearly $\delta_A(A_1^{(d)}) \subset \underline{end}(A) \otimes A_1^{(d)}$, it is induced in view of Lemma 6 by a morphism

$$\sigma^{(d)} : \underline{end}(A^{(d)}) \to \underline{end}(A)^{(d)} \to \underline{end}(A)$$

which is the third version of the quantum symmetric power (see the discussion in §2.10).

6. ADDING MISSING RELATIONS

1. Example. Start with the commutative polynomial algebra

$$A = k\langle \tilde{x}_1, \ldots, \tilde{x}_n \rangle / (\tilde{x}_i \tilde{x}_j - \tilde{x}_j \tilde{x}_i).$$

Then

$$A^! = k\langle \tilde{x}_1, \ldots, \tilde{x}_n \rangle / (\tilde{x}^k \tilde{x}^\ell + \tilde{x}^\ell \tilde{x}^k),$$

and

$$\underline{\mathrm{end}}\,(A) = k\langle \tilde{z}_i^k \rangle / (r_{(ij)}^{(k\ell)}),$$

$$\begin{aligned} r_{(ij)}^{(k\ell)} &= S_{(23)}((\tilde{x}^k \tilde{x}^\ell + \tilde{x}^\ell \tilde{x}^k)(\tilde{x}_i \tilde{x}_j - \tilde{x}_j \tilde{x}_i)) = \\ &= \tilde{z}_i^k \tilde{z}_j^\ell - \tilde{z}_j^k \tilde{z}_i^\ell + \tilde{z}_i^\ell \tilde{z}_j^k - \tilde{z}_j^\ell \tilde{z}_i^k = \\ &= [\tilde{z}_i^k, \tilde{z}_j^\ell] + [\tilde{z}_i^\ell, \tilde{z}_j^k] \end{aligned} \tag{1}$$

In all, we have $\frac{n(n-1)}{2} \cdot \frac{n(n+1)}{2}$ relations for n^2 entries z_i^k and, for $n > 1$, end (A) is highly noncommutative. But if we add $\frac{n^2(n^2-1)}{4}$ more relations, requiring (1) for the transposed matrix Z^t, we shall get $k[z_i^j]$, the common function ring of the matrix space. In fact, (1) means that for an arbitrary 2×2 submatrix of Z, the commutators of its diagonals mod R have opposite signs:

$$\begin{array}{c|cc} & k & \ell \\ \hline i & \bullet & \oplus \\ j & \oplus & \bullet \end{array}$$

But in the transposed matrix, just one of the commutators changes sign. Therefore, all of them must vanish.

2. General case. In the proof of Theorem 5.10 we have seen that the relations r for the transposed matrix essentially coincide with the relations $\check{r}$ defining end $(A^!)$. Therefore we may combine the generators $z_i^k, \check{z}_i^k$ and relations $r, \check{r}$, $z_i^k - \check{z}_i^k$ to achieve the necessary effect. However, the result may depend on the choice of coordinates x_i in A. Looking at the coordinate change formulas §5, (9) we see that the relations $Z - \check{Z}$ are invariant (up to a linear transformation) only with respect to the orthogonal coordinate changes $x' = Ux$, $UU^t = E$. This means that we implicitly fix an orthogonal form $g = x_1^2 + \ldots + x_n^2 \in S^2 A_1$, or identify A_1 and $A_1^!$ in a symmetric way. (Hence, e.g., $GL_q(2)$ from §1 is a 'cryptorthogonal" group! Of course, instead of g, A_1 may for some reason be equipped with a marked basis). Formally, we write

$$\underline{e}(A, g) = k\langle \tilde{z}_i^k, \check{z}_i^k \rangle / (r(\tilde{Z}), \check{r}(\check{Z}), \tilde{Z} - \check{Z})$$

$$\cong k\langle \tilde{z}_i^k \rangle / (r(\tilde{Z}), \check{r}(\tilde{Z}))$$

$$\cong k\langle \check{z}_i^k \rangle / (r(\check{Z}), \check{r}(\check{Z})).$$

3. Theorem.

a) The diagonal map $\tilde{\Delta}: \tilde{Z} \to \tilde{Z} \otimes \tilde{Z}$, $\check{Z} \to \check{Z} \otimes \check{Z}$ descends to $e(A, g)$; the maps $\delta(x_i) = \sum \tilde{z}_i^k \otimes \tilde{x}_k$, $\delta(x^i) = \sum \tilde{z}_i^k \otimes \tilde{x}^k$ define on A and $A^!$ the structures of left $e(A, g)$ – comodules.

b) The transposition map $r: \tilde{Z} \to \tilde{Z}^t$ defines an isomorphism of bialgebras $(\underline{e}(A, g), m, \Delta) \xrightarrow{\sim} (\underline{e}(A, g), m, \Delta^{op})$.

Proof. Everything follows from the previous discussion. In particular, the entries of $r(\tilde{Z}), \check{r}(\check{Z})$ and $\tilde{Z}-\check{Z}$ generate a coideal w.r.t. $\tilde{\Delta}$ in view of (7), §5, and

$$\tilde{\Delta}(\tilde{Z}-\check{Z}) = \tilde{Z}\otimes(\tilde{Z}-\check{Z}) + (\tilde{Z}-\check{Z})\otimes\check{Z}.$$

We leave the rest to the reader. ■

Of course, this construction is specially interesting if $R(A)$ bears some relation to g. We shall consider three examples.

4. Quantum conformal group. Let $A = k\langle x_i\rangle/(\sum x_i^2)$, i.e., $R(A) = kg, n \geq 2$. With the notation of §5, $m = 1, c_1^{ij} = \delta^{ij}$. Then (3), §5 shows that (c_{ij}^{β}) form a basis of trace-free matrices.

Hence $A^!$ can be defined by the relations

$$x^i x^j = 0 \qquad \text{for } i \neq j$$

$$(x^i)^2 = (x^j)^2 \quad \text{for all } i, j$$

In particular, $\dim A_1^! = n, \dim A_2^! = 1, \dim A_{\geq 3}^! = 0$ if $n \geq 2$ (and $A^! = k[x^1]$ if $n = 1$).

Furthermore, $\underline{end}(A)$ is defined by the relations

$$\sum_k z_k^i z_k^j = 0 \qquad \text{for } i \neq j,$$

$$(z_i^i)^2 = (z_j^j)^2 \quad \text{for all } i, j,$$

or

$$Z^t Z = \text{scalar}$$

for short. Finally

$$\underline{e}(A,g) = k(z_i^k)/(ZZ^t = \text{scalar},\ Z^tZ = \text{scalar}).$$

5. Case $R(A) = R(A)^{\perp}$ w.r.t. to g In this case $\tilde{g}: A_1 \to A_1^* = (A^!)_1$ induces an isomorphism $A \tilde{\to} A^!$ and therefore an isomorphism $\underline{end}(A) \tilde{\to} \underline{end}(A^!)$: $\tilde{Z} \to \check{Z} (\text{mod } R)$. This means that $\underline{e}(A,g) = \underline{end}(A)$.

Ezra Getzler has shown me a remarkable example of such a self-dual quadratic algebra, or rather of a bundle of such algebras over a riemannian space. It is a graded bundle associated with a filtration of the sheaf of differential operators acting upon a spinor bundle. (Actually, in this way one gets self-dual quadratic super algebras:. cf. §§11, 12).

6. Case $R(A) \oplus R(A)^{\perp} = A_1^{\otimes 2}$, $\dim R(A) = \frac{n(n-1)}{2}$. This case can be close to the commutative one, since the dimension of the space of quadratic relations for A(resp $A^!$, resp. $e(A,g)$) is the same as for $k[x_1,\ldots,x_n]$ (resp. Grassmann algebra of n variables, resp. $k[z_i^k]$). The deformation of the polynomial algebras, considered by Drinfeld [D1], belong to this class.

7. "Pseudosymmetric" quantum spaces. Consider an operator $R: A_1 \otimes A_1 \to A_1 \otimes A_1: \tilde{x}_i \otimes \tilde{x}_j \to \sum_{k\ell} r_{ij}^{k\ell} \tilde{x}_k \otimes \tilde{x}_\ell$. Assume that the following conditions are verified:

i) $r_{(ij)} := \sum_{pq}(\delta_{ij}^{pq} - r_{ij}^{pq})\tilde{x}_p \otimes \tilde{x}_q$, for $1 \leq i < j \leq n$ are linearly independent.

ii) $r^{(k\ell)} := \sum_{st}(\delta_{st}^{k\ell} + r_{st}^{k\ell})\tilde{x}^s \otimes \tilde{x}^t$, for $1 \leq k \leq \ell \leq n$ are linearly independent.

iii) $(R^2)_{ij}^{k\ell} := \sum_{pq} r_{ij}^{pq} r_{pq}^{k\ell} = \delta_{ij}^{k\ell}$, for $i<j$ and $k \leq \ell$.

Denote by A the quantum space defined by $r_{(ij)}$ and $g = \sum x_i^2$. Let $\tilde{Z} \odot \tilde{Z}$ be the $n^2 \times n^2$-matrix (usually denoted by the tensor product sign)

$$(\tilde{Z} \odot \tilde{Z})_{ij}^{k\ell} = \tilde{z}_i^k \otimes \tilde{z}_j^\ell$$

and $(R)_{ij}^{k\ell} = r_{ij}^{k\ell}$. Then we have:

8. Proposition.

a) $A^!$ is defined by $r^{(k\ell)}$, and <u>end</u> (A) is defined by (the entries of) $(I-R)(\tilde{Z} \odot \tilde{Z})(I+R)$.

b) If, in addition, $R^t = R$, $R^2 = 1$ then $\underline{e}(A, g)$ is defined by

$$R(\tilde{Z} \odot \tilde{Z}) - (\tilde{Z} \odot \tilde{Z})R \qquad (2)$$

Proof.

a) Since $\dim(\oplus k r_{(ij)}) = \frac{n(n-1)}{2}$, $\dim(\oplus k r^{(k\ell)}) = \frac{n(n+1)}{2}$, to check that $A^! = k\langle \tilde{x}^i \rangle / (r^{(k\ell)})$ it suffices to prove that $\langle r^{(k\ell)} | r_{(ij)} \rangle = 0$. In fact, using (iii):

$$\langle r^{(k\ell)} | r_{(ij)} \rangle = \sum_{s,t}(\delta_{st}^{k\ell} + r_{st}^{k\ell})(\delta_{ij}^{st} - r_{ij}^{st}) = (I-R)(I+R)|_{ij}^{k\ell} =$$

$$= \delta_{ij}^{k\ell} - (R^2)_{ij}^{k\ell} = 0.$$

Now, end (A) is defined by the relations

$$r^{(k\ell)}_{(ij)} = S_{(23)}(r^{(k\ell)} \otimes r_{(ij)}) = \sum_{pqst} (\delta^{k\ell}_{st} + r^{k\ell}_{st})(\delta^{pq}_{ij} - r^{pq}_{ij})\tilde{z}_p^{\ s} \otimes \tilde{z}_q^{\ t} =$$

$$= (I-R)\tilde{Z} \odot \tilde{Z}(I+R) \qquad (3)$$

b) Writing (3) for $\tilde{Z}^t$ instead of $\tilde{Z}$ and transposing, we find

$$(I+R^t)(\tilde{Z} \odot \tilde{Z})(I-R^t). \qquad (4)$$

If $R^t = R$, (3) and (4) together are equivalent to

$$\{R(\tilde{Z} \odot \tilde{Z}) - (\tilde{Z} \odot \tilde{Z})R,\ \tilde{Z} \odot \tilde{Z} - R(\tilde{Z} \odot \tilde{Z})R\} \qquad (5)$$

which boils down to (2) for $R^2 = 1$. ■

Remark. One sees that if $R^t = -R$ then $\underline{e}(A, g)$ = end (A).

9. Proposition. For an arbitrary operator $R: A_1^{\otimes 2} \to A_1^{\otimes 2}$, the matrix relation (2) defines a coideal w.r.t. $\Delta(\tilde{Z}) = \tilde{Z} \otimes \tilde{Z}$ and hence defines a new bialgebra.

Proof. One easily sees that $\Delta(\tilde{Z} \odot \tilde{Z}) = S_{(23)}[(\tilde{Z} \odot \tilde{Z}) \otimes (\tilde{Z} \odot \tilde{Z})]$. Therefore if R commutes with $Z \odot Z$, it also commutes with $\Delta[Z \odot Z]$. ■

Remark. In the QIST method, (2) is used to define a quantum group by means of a (weak) Yang-Baxter operator R cf. §11.

7. FROM SEMIGROUPS TO GROUPS

1. Motivation. In this section we show that for quantum semigroups like $E = \underline{end}(A)$, $\underline{e}(A,g)$ there exists a universal map $\gamma: E \to H$ into a Hopf algebra H. In view of proposition 2.7, γ makes all multiplicative matrices in E invertible. Hence a natural idea is to add formally the necessary inverse matrices. The following construction suffices to treat $\underline{end}(A)$ and $\underline{e}(A,g)$.

2. Construction. Let E be a bialgebra generated by entries of a multiplicative matrix Z. Let $E = k\langle\tilde{Z}\rangle/R_0$, where R_0 is the ideal of relations among entries of Z, $Z = \tilde{Z} \bmod R_0$.

Let $\tilde{Z}_0 = \tilde{Z}$ and introduce a series of matrices $\tilde{Z}_1, \tilde{Z}_2, \ldots$ in such a way that all entries of $\tilde{Z}_0, \tilde{Z}_1, \ldots,$ generate a free associative algebra. Denote by H the factor algebra of $\tilde{H} = k\langle\tilde{Z}_0, \tilde{Z}_1, \tilde{Z}_2, \ldots\rangle$ quotiented by the ideal generated by the following relations.†

$$R_k = \begin{cases} \text{elements of } R_0 \text{ written for } \tilde{Z}_k \text{ instead of } \tilde{Z} & \text{for } k \equiv 0 \pmod 2 \\ \text{\_\_\_\_ '' \_\_\_\_ } R_0^{op} \text{ \_\_\_\_ '' \_\_\_\_ '' \_\_\_\_} & \text{for } k \equiv 1 \pmod 2 \end{cases} \quad (1)$$

$$\tilde{Z}_k\tilde{Z}_{k+1} - I, \quad \tilde{Z}_{k+1}\tilde{Z}_k - I \quad \text{for } k \equiv 0 \pmod 2 \quad (2)$$

$$\tilde{Z}_k^t\tilde{Z}_{k+1}^t - I, \quad \tilde{Z}_{k+1}^t\tilde{Z}_k^t - I \quad \text{for } k \equiv 1 \pmod 2 \quad (3)$$

Define $\gamma: E \to H$ by $\gamma(Z) = \tilde{Z}_0 \bmod R$.

---

†) Here and henceforth $k\langle\hat{Z}_0, \hat{Z}_1, \ldots\rangle$ and similar expressions denote the algebra freely generated by the matrix entries.

3. Theorem.

a) Relations (1) – (3) generate a coideal $\tilde{R}$ with respect to $\tilde{\Delta}:\tilde{H}\to\tilde{H}\otimes\tilde{H}$,

$$\tilde{\Delta}(\tilde{Z}_k) = \begin{cases} \tilde{Z}_k\otimes\tilde{Z}_k & \text{for } k\equiv 0 \pmod 2, \\ (\tilde{Z}_k^t\otimes\tilde{Z}_k^t)^t & \text{for } k\equiv 1 \pmod 2 \end{cases}$$

so that $\tilde{\Delta}$ induces a comultiplication $\Delta:H\to H\otimes H$. Together with $\varepsilon(\tilde{Z}_k \bmod R) = 1$ this makes H a bialgebra and $\gamma:E\to H$ a bialgebra morphism.

b) The map $\tilde{i}:\tilde{H}\to\tilde{H}$, defined by $\tilde{i}(\tilde{Z}_k) = \tilde{Z}_{k+1}$, $\tilde{i}(fg) = \tilde{i}(g)\tilde{i}(f)$, verifies $\tilde{i}(\tilde{R}) \subset \tilde{R}$. Hence it induces a linear map $i:H\to H$ which is an antipode of (H, m, Δ).

c) For any bialgebra morphism $\gamma':E\to H'$, where H' is a Hopf algebra, there exists a unique Hopf algebra morphism $\beta:H\to H'$ such that $\gamma' = \beta\circ\gamma$.

Proof.

a) Since E is a coalgebra w.r.t. $\Delta(Z) = Z\otimes Z$, we have $\tilde{\Delta}(R_0) \subset k\langle\tilde{Z}_0\rangle \otimes R_0 + R_0\otimes k\langle\tilde{Z}_0\rangle \subset \tilde{H}\otimes R_0 + R_0\otimes\tilde{H}$. It follows that the ideal generated by R_0 (and by all $R_k, k\equiv 0 \pmod 2$) is a coideal. One treats similarly $R_k, k\equiv 1 \pmod 2$, using $(E, m_E^{op}, \Delta_E^{op})$ instead of (E, m, Δ).

To treat (2) and (3), let us consider the following identity: let A, B be two matrices in H with $\Delta(A) = A\otimes A$, $\Delta(B) = (B^t\otimes B^t)^t$. Then

$$\Delta(AB)_i^k = [(A\otimes A)(B^t\otimes B^t)^t]_i^k = \sum_j (A\otimes A)_i^j (B^t\otimes B^t)_k^j =$$

$$= \sum_{jrs} (a_i^r \otimes a_r^j)(b_s^k \otimes b_j^s) = \sum_{rs} a_i^r b_s^k \otimes (AB)_r^s.$$

Hence

$$\Delta(AB-I)_i^k = \sum_{rs} a_i^r b_s^k \otimes [(AB)_r^s - \delta_r^s] + (\sum_r a_i^r b_r^k - \delta_i^k)\otimes 1,$$

and one sees that entries of AB–I generate a coideal. This disposes of the first relations in (2), (3). The second ones can be treated similarly. Obviously $\tilde{\varepsilon}: \tilde{Z}_k \to I$ descends to a counit.

b) Clearly, $\tilde{i}(R) \subset \tilde{R}$. Let $i: H \to H$ be the induced linear map. We must check that for each $u \in H$

$$m\circ(i\otimes id)\circ\Delta(u) = m\circ(id\otimes i)\circ\Delta(u) = \eta\,\varepsilon(u) \qquad (4)$$

(cf. §2.2). From the definition of $\tilde{\Delta}$ and $\tilde{\varepsilon}$ it follows that this is true for all those u that are entries of $Z_k = \tilde{Z}_k \bmod \tilde{R}$. Since they generate H as a ring, it suffices to check that if (4) is true for u and v it is true for uv. In fact

$$m\circ(i\otimes id)\circ\Delta(u)\Delta(v) = m\circ(i\otimes id)(\sum_k u'_k\otimes u''_k)(\sum_\ell v'_\ell\otimes v''_\ell) =$$

$$= m(\sum_{k\ell} i(v'_\ell)i(u'_k)\otimes u''_k v''_\ell) =$$

$$= \sum_\ell i(v'_\ell)(\sum_k i(u'_k)u''_k)v''_\ell = \eta\,\varepsilon(u)(\sum_\ell i(v'_\ell)v''_\ell) = \eta\,\varepsilon(u)\,\eta\,\varepsilon(v)$$

c) Let $\gamma': E \to H'$ be a morphism into a Hopf algebra. Put $Y'_k = i^k_{H'}(\gamma'(Y))$. In view of Theorem 2.4 and Proposition 2.7 the entries of Y'_k verify relations (1) – (3). Hence $\tilde{B}: \tilde{Z}_k \to Y'_k$ induces an algebra morphism $\beta: H \to H'$ which is clearly compatible with Δ and i. ∎

4. Remarks.

a) This construction has several useful variations. First, one can construct a universal map $\overline{\gamma}: E \to \overline{H}$ in the class of Hopf algebras with bijective antipode. To this end, one has only to adjoin matrices $\tilde{Z}_k$ and relations (1) – (3) for all $k \in Z$.

Second, one can construct a universal map $\overline{\gamma}_d: E \to \overline{H}_d$ in the class of Hopf algebras with antipode i verifying $i^{2d} = id$. To do this, one should add to the previous relations $\tilde{Z}_k - \tilde{Z}_{k+2d}$ for all k.

Third, one can construct "formal quantum groups" corresponding to $H, \overline{H}, \overline{H}_d$. They are completions of these rings w.r.t. the ideal generated by entries of $Z_k - I$. Continuous dual Hopf algebras of these objects are quantum analogues of universal enveloping algebras. Some of these were studied by Drinfeld and Jimbo.

b) If E is a quadratic algebra, like $\underline{end}$ (A) or $\underline{e}(A, g)$, it is nice to have a quadratic version of $H, \overline{H}, \overline{H}_d$ in view of the good properties of QA. Since only relations (2) – (3) are not quadratic, we may add central elements t_k and homogenize (2) – (3) to $\tilde{Z}_k \tilde{Z}_{k+1} - t^2 I$ etc. Clearly, $\tilde{\Delta}(t_k) = t_k \otimes t_k$ will do. We leave it to the reader to make explicit the universal properties of this construction.

5. (Co) representations. Clearly, $A \xrightarrow{\delta} E \otimes A \xrightarrow{\beta \otimes id} H \otimes A$ (for $E = \underline{end}$ (A), $\underline{e}$ (A, g)) is a corepresentation of the "Hopf envelope" H. Using the universal property of H, one can repeat everything that was said in §5.10 – 5.13 for H instead of E.

In particular, we have the following results: $\underline{gl}(A) :=$ Hopf envelope of $\underline{end}$ (A) acts upon

$$A \text{ via } \delta(x.) = Z_0 \otimes x. \quad (\text{or } Z_k \otimes x., \quad k \equiv 0 \pmod 2)$$

$$A^{!op} \text{ via } \delta(\check{x}.) = Z_1^t \otimes \overset{o}{\check{x}}. \quad (\text{or } Z_{k+1} \otimes \overset{o}{\check{x}}., k \equiv 0 \pmod 2)$$

In addition, $\underline{gl}(A, g) :=$ Hopf envelope of $\underline{End}$ (A, g) acts upon

$$A^! \text{ via } \delta(x.) = Z_0 \otimes x.,$$

$$A^{op} \text{ via } \delta(\check{x}.) = Z_1^t \otimes \check{x}..$$

8. FROBENIUS ALGEBRAS AND THE QUANTUM DETERMINANT

1. Definition. A quadratic algebra A is (a function algebra of) a Frobenius quantum space of dimension d (F.a. for short), if

a) $\dim A_d = 1$, $A_i = 0$ for $i > d$

b) For all j, the multiplication map $m : A_j \otimes A_{d-j} \to A_d$ is a perfect duality.

A is called a quantum grassmann algebra if, in addition,

c) $\dim A_i = \binom{d}{i}$. ■

2. Quantum determinant. Let E be a bialgebra, acting upon a Frobenius space A via $\delta : A \to E \otimes A$, $\delta(A_1) \subset E \otimes A_1$. Then $\delta(A_i) \subset E \otimes A_i$, and for $i = d$ we get an element $D = \mathrm{DET}(\delta) \in E$ defined by

$$\delta(A) = \mathrm{DET}(\delta) \otimes a \quad \text{for all } a \in A_d.$$

Clearly, D is multiplicative (§2.9) :

$$\delta(D) = D \otimes D, \qquad \varepsilon(D) = 1.$$

In the language of §5.13, D defines the d–th quantum symmetric power of the corepresentation A_1.

3. Quantum Cramer and Lagrange identities. Choose bases $\{X_i^{(j)}\}$ in A_j, $\{Y_k^{(\ell)}\}$ in A_ℓ in such a way that in such

$$X_i^{(j)} Y_k^{(d-j)} = \delta_{ik} a \tag{1}$$

for a fixed $a \in A_d \setminus \{0\}$. Define multiplicative matrices $Z^{(j)}$ and $V^{(\ell)}$ describing $\delta : A_k \to E \otimes A_k$ in these bases :

$$\delta(X_i^{(j)}) = \sum_k z_i^{(j)} \otimes X_\ell^{(j)}, \quad \delta(Y_i^{(j)}) = \sum_m v_i^{(j)m} \otimes Y_m^{(j)} \tag{2}$$

Applying δ to (1) and taking into account (2), we get

$$Z^{(j)} V^{(d-j)t} = \mathrm{DET}(\delta)I. \tag{3}$$

If A is a standard grassmann algebra and $X_i^{(j)}$, $Y_k^{(d-j)}$ are exterior monomials of a basis of A_1, then (3) gives the classical Cramer identities for $j = 1$ and Lagrange identities for general j (when E is commutative).

4. Theorem. Let H be a Hopf algebra acting on a Frobenius algebra A as in no. 2, D the determinant of this action. Then

a) D is invertible.

b) $H/(D-1)$ is a Hopf algebra, with Δ, i and ε induced from H.

Proof.

a) Since D is a matrix of a one–dimensional corepresentation it is invertible in view of Proposition 2.7 :

$$i(D) = D^{-1}$$

b) $D-1$ generates a coideal in H, since

$$\Delta(D-1) = D \otimes (D-1) + (D-1) \otimes 1.$$

This coideal is i–stable, since

$$i(D-1) = D^{-1} - 1 = D^{-1}(D-1).$$

Finally, $D-1 \in \mathrm{Ker}\,\varepsilon$. ■

5. General and special linear groups of a quantum space. Let A be a quadratic algebra. The Hopf envelopes of $\underline{end}(A)$ and $\underline{e}(A,g)$ constructed in §7 deserve to be called general linear groups of A : $\underline{g\ell}(A)$, $\underline{g\ell}(A,g)$. If, in addition, A is Frobenius, we can define special linear groups

$$\underline{s\ell}(A) = \underline{g\ell}(A)/(D-1), \qquad \underline{s\ell}(A,g) = \underline{g\ell}(A,g)/(D-1),$$

where D is the appropriate determinant.

6. Example. Let $A = k\langle x_1,\dots,x_n\rangle/(x_i^2;\, x_ix_j + qx_jx_i$ for $i<j)$, $q \in k^{\bullet}$. Clearly, A is n-dimensional Frobenius (in fact, even quantum grassmannian). Writing $\delta(x_i) = \Sigma z_i^k \otimes x_k$, $a = x_1,\dots,x_n \in A_n\setminus\{0\}$, we get the formula for calculating $DET(\delta)$:

$$\prod_{i=1}^{n}\Big(\sum_{k=1}^{n} z_i^k \otimes x_k\Big) = DET(\delta) \otimes \prod_{i=1}^{n} x_i ,$$

which gives

$$DET(\delta) = \sum_{s\in S_n} (-q)^{-\ell(s)} z_1^{s(1)},\dots,z_n^{s(n)}$$

(cf. [FRT]).

7. Example. Let $B = k\langle x_1,\dots,x_n\rangle/(x_ix_j$ for $i \neq j$; $(x_i)^2 = (x_j)^2$ for all $i, j)$, $n \geq 2$ (cf. §6.4). This is a two-dimensional Frobenius algebra. Taking $a = (x^i)^2$ we obtain for $DET(\delta)$ the formula

$$DET(\delta) = \sum_{k=1}^{n} (z_i^k)^2 \quad \text{for each } i = 1,\dots,n.$$

One can similarly obtain two-dimensional Frobenius algebras with quadratic determinant starting from a 2d-dimensional F.a. A and putting $B = A^{(d)} = k \oplus A_d \oplus A_{2d}$. Cf. also Gurevich's construction of a Yang-Baxter symmetric algebra which is a 2-dim F.a.

8. Example (Drinfeld [D2]). Let k be the quotient field of a local ring L. Consider the deformed anticommutation relations

$$x^k x^\ell + x^\ell x^k = \sum_{ij} c_{ij}^{k\ell} x^i x^j \qquad (4)$$

where $c_{ij}^{k\ell}$ belong to the maximal ideal of L and

$$c_{ij}^{k\ell} = c_{ij}^{\ell k} = -c_{ji}^{k\ell}.$$

Intuitively, this means that the corresponding algebra is a small deformation of the exterior algebra, at least from the point of view of its relations. Clearly, for $c_{ij}^{k\ell} = 0$ (4) defines the usual grassmann algebra. In [D2] Drinfeld calculated conditions on $c_{ij}^{k\ell}$ which must be verified in order that (4) define a quantum grassmann algebra over L (and a fortiori, over k). Let

$$b_{i_1 i_2 i_3}^{j_1 j_2 j_3} = \sum_j c_{i_1 j}^{j_1 j_2} c_{i_2 i_3}^{j j_3}, \quad b = \left(b_{i_1 i_2 i_3}^{j_1 j_2 j_3} \right), \quad a = b\left(1 - \frac{b}{3}\right)^{-1}.$$

Then

$$\text{(4) gives a q. g. a. over } L \Leftrightarrow \text{alt(i) sym(j) } a_{i_1 i_2 i_3}^{j_1 j_2 j_3} = 0. \qquad (5)$$

If (5) is true then A is n–dimensional and we get DET(δ) in end(A). However, it is impossible to calculate it explicitly as in example 5, since we do not know the coefficients d(s) in the expressions

$$x_{s(1)},\ldots,x_{s(n)} = d(s)x_1,\ldots,x_n \text{ mod } R.$$

9. Example. $M_q(2)^!$ is a Frobenius quantum space of dimension 4. In fact, it is even grassmannian. We leave to the reader the calculation of the corresponding DET.

9. KOSZUL COMPLEXES AND THE GROWTH RATE OF QUADRATIC ALGEBRAS

1. Lemma. For a quadratic algebra A, as in §5.1, let

$$\xi_A = \xi = \sum_{i=1}^{n} x^i \otimes x_i \in A^! \circ A \subset A^! \otimes A.$$

Then $\xi^2 = 0$,

Proof. We have

$$\xi^2 = \sum_{ik} \tilde{x}^i \tilde{x}^k \otimes \tilde{x}_i \tilde{x}_k \ (\mathrm{mod}\ R) = \sum_{\lambda} X^{\lambda} \otimes X_{\lambda} \ (\mathrm{mod}\ R),$$

where $R = R(A^! \circ A)$ and (X^{λ}), (X_{μ}) are arbitrary dual bases of $\tilde{A}_2$. Take $X_{\lambda} = \{r_{\alpha}, s_{\beta}\}$, $X^{\lambda} = \{s^{\alpha}, r^{\beta}\}$ as in the proof of Lemma 5.5. Then

$$\xi^2 = \sum_{\alpha} s^{\alpha} \otimes r_{\alpha} + \sum_{\beta} r^{\beta} \otimes s_{\beta} \ \mathrm{mod}\ R = 0. \quad \blacksquare$$

Remarks.

a) Theorem 4.2 gives a conceptual explanation of this lemma. In fact, $\mathrm{Hom}(k[\varepsilon] \bullet A, A) = \mathrm{Hom}(k[\varepsilon], A^! \circ A)$ and $k[\varepsilon] \bullet A = A$. Under this correspondence, id_A goes to $k[\varepsilon] \to A^! \circ A$, $\varepsilon \to \xi_A$.

b) More generally, for any morphism $f: B \to A$ in a QA, we obtain an element $\xi_f \in B^! \circ A$ with $\xi_f^2 = 0$. It corresponds to f under the isomorphism of Theorem 4.2: $\mathrm{Hom}(k[\varepsilon] \circ B, A) = \mathrm{Hom}(k[\varepsilon] \circ B^! = A)$. We have

$$\xi_A = \xi_{id_A};\ \xi_f = (id \circ f)(\xi_B).$$

2. Koszul complexes $L^{\bullet}$. For any ring B and $b \in B$ denote by $\ell(b)$ (resp. $r(b)$) the left (resp right) multiplication by b in B (or in a B–module). For a morphism f in QA we can define two complexes

$L^{\bullet}(f) = (B^{!} \otimes A,\ r(\xi_f)$ or $\ell(\xi_f))$. Since $\xi_f \in B^{!}_1 \otimes A_1$, these are direct sums of subcomplexes

$$L^{p}(f) = \bigoplus_{b-a=p} B^{!}_b \otimes A_a,\ r(\xi_f) \text{ or } \ell(\xi_f)),\ p \in \mathbf{Z}.$$

We shall write $L^{p}(A) = L^{p}(id_A)$, and set $L^{p,a}(f) = B^{!}_b \otimes A,\ b = p+a$.

3. Proposition.

a) Suppose that at least one of the rings $B^{!}$, A is finite dimensional. Then all complexes $L^{n}(f)$ are finite, and we can define the formal series

$$\chi^{L}_f(t) = \sum_n \chi(L^{n}(f))t^{n}, \quad \chi(L^{n}(f)) = \sum_A (-1)^{a} \dim H^{a}(L^{n}(f)).$$

Let $P_A(t) = \Sigma \dim A_i t^{i}$. We have

$$P_{B^{!}}(t)\, P_A(-t^{-1}) = \chi^{L}_f(t). \tag{1}$$

b) With the same assumptions, let $p = \max\{b \mid B^{!}_b \neq \{0\}\}$, $q = \max\{a \mid A_a \neq \{0\}\}$. Then, for $H^{a,n} = H^{a}(L^{n}(f))$,

$$H^{p,0} = B^{!}_p,\ H^{-q,0} = A_q \tag{2}$$

In particular, if $L^{\bullet}(f)$ is acyclic everywhere except for (2), for either $r(\xi_f)$ or $\ell(\xi_f)$, then

$$P_{B^{!}}(t)\, P_A(-t^{-1}) = t^{p} + t^{-q}(-1)^{q}$$

(where t^{p} (resp. t^{-q}) should be interpreted as zero if p (resp. q) is not defined).

Proof. It is quite standard : we have

$$\chi(L^n(f)) = \sum_{b-a=n} \dim B^!_b \dim A_a (-1)^a.$$

Multiplying by t^n and adding up we get (1). The rest is clear. ■

4. Homological determinant. Looking at (2) we see, that the determinant of a bialgebra E acting upon a Frobenius algebras A is defined by its corepresentation on one of the cohomology groups of L(A). In fact, one can define the corepresentation of E on this cohomology, assuming that $R(A^!) = R(A^!)^{op}$ and E is a Hopf algebra. Then if any one–dimensional cohomology group of $L^*(A)$ happens to exist, it gives rise to a corepresentation of E, which can be called a homological determinant.

We shall prove this for the universal situation $E = \underline{g\ell}(A)$. Our construction is motivated by the cohomological treatment of Berezinian in superalgebras (See e.g. Yu. I. Manin, Gauge fields and complex geometry, Springer, 1988). In fact, this construction gives a quantum Berezinian for a class of "superFrobenius" quantum supergroups.

5. Proposition. Assume that $R(A^!) = R(A^!)^{op}$. Then the natural action of $\underline{g\ell}(A)$ upon A and $A^{!op} = A^!$ defined in §7.5 induces a corepresentation

$$H^*(L(A)) \to \underline{g\ell}(A) \otimes H^*(L(A)).$$

Proof. Clearly, $\underline{g\ell}(A)$ acts upon $A^! \otimes A$ via

$$A^! \otimes A \xrightarrow{\delta_{A^!} \otimes \delta_A} \underline{g\ell}(A) \otimes A^! \otimes \underline{g\ell}(A) \otimes A \xrightarrow{(m_{\underline{g\ell}(A)} \otimes id)S_{(23)}} \underline{g\ell}(A) \otimes A^! \otimes A.$$

Let us calculate the image of ξ_A in the notation of §7.5 :

$$\sum_i \check{x}_i \otimes x_i = \xi_A \to \sum_{ijk} (Z_1^t)_i^j \otimes \check{x}_j \otimes (Z_0)_i^k \otimes x_k$$

$$\to \sum_{ijk} (Z_1 Z_0)_j^k \otimes \check{x}_j \otimes x_k = 1 \otimes \xi_A$$

since $Z_1 Z_0 = I$ in view of (2), §7. Therefore we get a morphism of complexes

$$(A^! \otimes A,\ r(\xi) \text{ or } \ell(\xi)) \to (\underline{g\ell}(A) \otimes A^! \otimes A,\ r(1 \otimes \xi) \text{ or } \ell(1 \otimes \xi)),$$

which induces our corepresentation. ■

Remark. I do not know whether one really needs the condition $R(A^!) = R(A^!)^{op}$. Perhaps, one can omit it by slightly changing our construction.

6. Koszul complexes $K^\bullet$: a construction. We start with the following simple construction. Suppose we are given a diagram of linear spaces (or objects of an abelian category) :

$$\begin{array}{ccccc} E & \subset & F & \subset & G \\ \cup & & \cup & & \cup \\ E' & \subset & F' & \subset & G' \end{array} \qquad (3)$$

which induces natural maps $E/E' \xrightarrow{\alpha} F/F' \xrightarrow{\beta} G/G'$. Then :

a) $\beta\alpha = 0 \Leftrightarrow E \subset G'$.

b) If a) is true then $H := \mathrm{Ker}\beta/\mathrm{Im}\alpha \cong (F \cap G')/(E+F')$.

Now let $A^{!*} = \bigotimes_a A_a^{!*}$ where star means linear dualization, and consider the following diagram of the type (3).

$$\begin{array}{ccccc} A^{!*}_{a+1}\otimes A_1^{\otimes b-1} & \hookrightarrow & A^{!*}_{a}\otimes A_1^{\otimes b} & \hookrightarrow & A^{!*}_{a-1}\otimes A_1^{\otimes b+1} \\ \cup & & \cup & & \cup \\ A^{!*}_{a+1}\otimes R_{b-1}(A) & \hookrightarrow & A^{!*}_{a}\otimes R_{b}(A) & \hookrightarrow & A^{!*}_{a-1}\otimes R_{b+1}(A). \end{array} \tag{4}$$

In order to explain the horizontal inclusion maps recall that

$$A^{!}_{a} = A_1^{*\otimes a} / \sum_{i=0}^{a-2} A_1^{*\otimes i} \otimes R(A)^{\perp} \otimes A_1^{*a-2-i}$$

and hence we can identify

$$A^{!*}_{a} = \bigcap_{i=0}^{a-2} A_1^{\otimes i} \otimes R(A) \otimes A_1^{a-2-i} \subset A_1^{\otimes a}.$$

Therefore the subspaces

$$A^{!*}_{a}\otimes A_1^{\otimes b} = \bigcap_{i=0}^{a-2} A_1^{\otimes i} \otimes R(A) \otimes A_1^{a+b-2-i} \subset A_1^{\otimes (a+b)}$$

clearly form an increasing filtration of $A_1^{\otimes(a+b)}$ when a decreases. As for the lower line, we have similarly

$$\begin{aligned} & A^{!*}_{a}\otimes R_b(A) = \\ & = (A^{!*}_{a}\otimes A_1^{\otimes b}) \cap (A_1^{\otimes a}\otimes R_b(A)) = \\ & = (\bigcap_{i=0}^{a-2} A_1^{\otimes i} \otimes R(A) \otimes A_1^{a+b-2-i}) \cap (\sum_{j=a}^{a+b-2} A_1^{\otimes j} \otimes R(A) \otimes A_1^{a+b-2-j}) \end{aligned} \tag{5}$$

so that these subspaces also increase with decreasing a. Finally, to check that (4) generates a complex we must convince ourselves that

$$A^{!*}_{a+1}\otimes A_1^{\otimes b-1} \subset A^{!*}_{a-1}\otimes R_{b+1}(A)$$

i.e., that

$$\bigcap_{i=0}^{a-1} A_1^{\otimes i} \otimes R(A) \otimes A_1^{\otimes a+b-2-i} \subset$$

$$\subset \Big(\bigcap_{i=0}^{a-3} A_1^{\otimes i} \otimes R(A) \otimes A_1^{\otimes a+b-2-i}\Big) \cap \Big(\sum_{j=a-1}^{a+b-2} A_1^{\otimes i} \otimes R(A) \otimes A_1^{\otimes a+b-2-j}\Big).$$

(Use (5)). But this inclusion is clear since $\bigcap_{i=0}^{a-1} \subset \bigcap_{i=0}^{a-3}$ and $\bigcap_{i=0}^{a-1} \subset$ (a−1)–th member of $\cap$ = (a−1)–th member of Σ.

Since finally

$$A_a^{!*} \otimes A_1^{\otimes b} / A_a^{!*} \otimes R_b(A) = A_a^{!*} \otimes A_b,$$

we get a new Koszul complex

$$K^{a+b,\bullet}(A) : \ \ldots \to A_{a+1}^{!*} \otimes A_{b-1} \to \underset{\substack{\uparrow \\ \text{b–th place}}}{A_a^{!*} \otimes A_b} \to A_{a-1}^{!*} \otimes A_{b+1} \to \ldots$$

Since we neglected in the previous discussion several border effects let us describe explicitly the lower complexes :

$$K^{0,\bullet} : \quad \ldots \to 0 \to k \to 0 \to \ldots \tag{6}$$

$$K^{1,\bullet} : \quad \ldots \to 0 \to \underset{(0)}{A_1^{!*} = A_1} \overset{id}{\to} \underset{(1)}{A_1} \to 0 \to \ldots \tag{7}$$

$$K^{2,\bullet} : \quad \ldots \to 0 \to \underset{(0)}{A_2^{!*} = R(A)} \to \underset{(1)}{A_1^{!*} \otimes A_1 = A_1^{\otimes 2}} \to \underset{(2)}{A_0^{!*} \otimes A_2} \to 0 \tag{8}$$

We see that $K^{a+b,\bullet}$, unlike $L^{a+b,\bullet}$, are always *finite*. If all complexes $K^{p,\bullet}$, with the exception of $p=0$, are acyclic, A is called a Koszul algebra. Many nice properties of Koszul algebras are known : cf. [P], [Lö], [BF].

Writing $\chi^K_A(t) = \sum_n \chi(K^{n,\bullet}(A))t^n$ we get, as in no. 3 :

7. Proposition. $P_{A^!}(t)P_A(-t) = \chi^K_A(t)$. In particular, if A is Koszul, then

$$P_{A^!}(t)P_A(-t) = 1. \quad ■$$

8. Complexes $K^\bullet(f)$. Now let $f : B \to A$ be a morphism of QA's. It induces morphisms $(id\otimes f) : B^{!*}\otimes B \to B^{!*}\otimes A$ and $B^{!*}\otimes A \xrightarrow{(f^{!*}\otimes id)} A^{!*}\otimes A$. Imitating the construction of no. 6 one can construct a differential on $B^{!*}\otimes A$, depending on f, in such a way that both these maps become complex morphisms. This complex will be denoted $K^\bullet(f)$. It can be obtained by partial dualization of $L^\bullet(f)$.

Suppose that $K^p(f)$ is acyclic for $p > 0$. Looking at (7), (8) one sees that in this case f is an isomorphism and A, B are Koszul algebras.

Now we shall show how to use the Koszul complexes to estimate the "size" of some quadratic algebras.

For a graded algebra A, we shall sometimes write $P(A,t)$ instead of $P_A(t)$. For two formal series $f_i(t) = \sum_j a_{ij}t^j$, $i = 1,2$, we define $(f_1 * f_2)(t) = \sum_j a_{1j}a_{2j}t^j$. With this notation, we have

9. Proposition.

a) Let A, B be Koszul. Then

$$P(\underline{hom}(B, A), t) = [P(B, -t) * P(A, -t)^{-1}]^{-1} \tag{9}$$

b) Let B be Koszul and either A or $B^!$ be finite dimensional. Then for a morphism $f: B \to A$

$$\chi_f^L(t) = P(A, -t^{-1})\, P(B, -t)^{-1}. \tag{10}$$

Proof.

a) The class of Kozsul algebras is stable with respect to ! and $\circ$ (cf. [BF]), and therefore with respect to $\bullet$. Hence, using Proposition 9.7 repeatedly, we find

$$P(\underline{hom}(B, A), t) = [P(B^! \bullet A, t) = P(B \circ A^!, -t)^{-1} =$$
$$= [P(B, -t) * P(A^!, -t)]^{-1} = [P(B, -t) * P(A, -t)^{-1}]^{-1}.$$

b) This follows from Propositions 9.3 and 9.7. ■

Remark. We shall use (9) below for B = A in order to calculate the degree of "hidden symmetry" of a quadratic algebra. On the other hand, (10) sometimes allows us to estimate a possible range of nontrivial cohomological representations of $\underline{end}(A)$ on $H^\bullet(L(A))$.

10. TABLE

	$[1 * 2]^{-1}$	$[2 \bullet 3]$	
A	$P(A, -t)$	$P(A, -t)^{-1}$	$P(A, -t^{-1})$
$S_n = k[x_1, \ldots, x_n]$	$\dfrac{1}{(1+t)^n} = \sum_{j\geq 0} \binom{j+n-1}{n-1} t^j (-1)^j$	$(1+t)^n = \sum_{k=0}^{n} \binom{n}{k} t^k$	$\dfrac{t^n}{(1+t)^n}$
$\Lambda_m = S_m^! = k\langle \xi_1, \ldots, \xi_m \rangle / [\xi_i, \xi_j]_+$	$(1-t)^m = \sum_{j=0}^{m} (-1)^j \binom{m}{j} t^j$	$\dfrac{1}{(1-t)^m} = \sum_{k\geq 0} \binom{k+m-1}{m-1} t^k$	$\dfrac{(1-t)^m}{(-t)^m}$
$E = k[x_1, \ldots, x_4]/(q_1, q_2)$ (complete intersection of two quadrics = elliptic curve)	$\dfrac{(1-t)^2}{(1+t)^2} = 1 + 4 \sum_{j=1}^{\infty} j t^j (-1)^j$	$\dfrac{(1+t)^2}{(1-t)^2} = 1 + 4 \sum_{k=1}^{\infty} k t^k$	$\dfrac{(t-t)^2}{(1+t)^2}$

Note in particular that

$$P(E, t) = P(S_2 \otimes \Lambda_2, t).$$

Note also that since E and $E^!$ are infinite dimensional we cannot calculate χ^L from the table. Is $L^\bullet$ essentially acyclic?

11. Calculations.

a) $\dim(\underline{end}(S_2)_i) = \frac{3^{i+1}-1}{2}$. In fact,

$$\frac{1}{(1+t)^2} * (1+t)^2 = 1-4t+3t^2 = (1-t)(1-3t),$$

$$P\,(\underline{end}(S_2), t) = \frac{3/2}{1-3t} - \frac{1/2}{1-t}.$$

b) $P\,(\underline{end}(E), t) = \frac{(1+t)^3}{1-13t+19t^2+t^3}$. The smallest root of the denominator is 0.08915047 ..., so that $\dim \underline{end}\,(E)_i$ grows like const. $(11.21690\,...)^i$. Note that $\dim E_1 = 4$ so that 16^i is the growth rate of the corresponding free matrix algebra.

12. Problem. How can one systematically calculate $\dim\,(\underline{end}\,(A, g)_i)$? I do not know of any criterion to ensure that $\underline{end}(A, g)$ be Koszul. Of course, in some cases one can find explicit bases in $\underline{end}(A, g)_i$. Consider, for example, the exceptional case $q^2 = -1$ for $E = \underline{end}(A_q^{2|0}, g)$ of §1 where the "flat" dependence of q breaks down. We can verify that in this case E has a basis consisting of monomials

$$a^\alpha\, b^\beta\, (cb)^\gamma c^\delta\, d^\varepsilon.$$

Counting the number of such monomials, one obtains the order of growth corresponding to that of a 5–dimensional commutative variety, while for $q^2 \neq -1$ it is 4–dimensional.

It is known that in the theory of Hecke algebras the values $q \in$ {roots of unity} generally play a special role. Perhaps, their exceptional properties would show in the structure of $\underline{end}(A_q^{n|0}, g)$ where $A_q^{n|0}$ is given by $x_i x_j = q^{-1} x_j x_i$ for $i < j$.

Generally speaking, one would expect that a Hecke algebra is a "quantized Weyl group" for $\underline{g\ell}(A_q^{n|0}, g)$. Can we make this precise and define quantum Weyl groups in a more general setting?

10. HOPF * – ALGEBRAS AND COMPACT MATRIX PSEUDOGROUPS

1. Hopf * – algebras. Let (H, m, Δ) be a Hopf **C**–algebra with a bijective antipode i. V.G. Drinfeld suggested defining a *–structure on H to be an antilinear map $j : E \to E$ with the following properties :

a) j is an isomorphism of algebras and an anti–isomorphism of coalgebras.

b) $j^2 = (ij)^2 = \mathrm{id}$.

There is a method of construction of Hopf * – algebras. Let E be a bialgebra over **C** generated by entries of a multiplicative matrix Z such that there is a **C**–isomorphism $\tau : (E, m, \Delta) \to (E, m, \Delta^{op})$ for which $\tau(Z) = Z^t$. Assume that in addition $R(E) \subset E_1^{\otimes 2}$ is a real subspace with respect to the real structure generated by tensor products of entries of Z. Then we can define $j_E = \{\tau$ followed by complex conjugation of coefficients$\}$. Clearly, j_E verifies a) and $j_E^2 = \mathrm{id}$.

2. Proposition. Let $\gamma : E \to H$ be a universal map of E into a Hopf **C**–algebra with bijective antipode i. Then there is a unique *–structure j on H such that $j(\gamma(Z)) = \gamma(Z)^t$.

Proof. Recall the explicit construction of (H, γ) in §7.2:

$$H = \mathbf{C}\langle \tilde{Z}_i \mid i = \dots -1, 0, 1, \dots \rangle / \tilde{R}$$

where $\tilde{R}$ is generated by the relations §7, (1)–(3) for all $k \in \mathbf{Z}$. Define an antilinear algebra morphism $\tilde{j} : \mathbf{C}\langle \tilde{Z}_i \rangle \to \mathbf{C}\langle \tilde{Z}_i \rangle$ by

$$\tilde{j}(\tilde{Z}_i) = \tilde{Z}_{-i}^{t}.$$

Clearly, $\tilde{j}(\tilde{R}) = \tilde{R}$, so that $\tilde{j}$ descends to H. Moreover, the antipode of H is induced by $\tilde{i} : \tilde{Z}_k \to \tilde{Z}_{k+1}$. Therefore one easily sees that $(\widetilde{i\,j})^2 = \mathrm{id}$:

$$\tilde{Z}_k \xrightarrow{\tilde{j}} \tilde{Z}_{-k}^{-t} \xrightarrow{\tilde{i}} \tilde{Z}_{-k+1}^{t} \xrightarrow{\tilde{j}} \tilde{Z}_{k-1} \xrightarrow{\tilde{i}} \tilde{Z}_k.$$

We leave uniqueness to the reader. ■

3. Example. This construction is applicable to $E = \underline{e}(A, g)$ (cf. §6.2) if $R(A)$ is a real subspace in $A_1^{\otimes 2}$ with respect to the basis $x_i \otimes x_j$ in which $g = \Sigma x_i^2$. This construction applied to $GL_q(2)$ after a completion gives the quantum SU(2) considered e.g. by Woronowicz [W1].

4. Compact matrix pseudogroups. Woronowicz [W2] calls a compact matrix pseudogroup the following data: (E, Z, Δ), where

a) E is a $\mathbf{C}^*$–algebra, Z is a square matrix whose entries generate a dense $\star$–subalgebra E of E.

b) $\Delta : E \to E \otimes E$ is a comultiplication such that $\Delta(Z) = Z \otimes Z$ and Δ is a morphism of $\mathbf{C}^*$–algebras.

c) E with the induced structure of bialgebra has a bijective antipode i (denoted κ by Woronowicz).

d) $i(i(a^*)^*) = a$ for all $a \in E$.

Let us discuss connections between this notion and the notion of a Hopf $\star$–algebra.

If a compact matrix pseudogroup (E, Z, Δ) is given then a $*$–structure on E in Drinfeld's sense can be defined by $j: a \to i^{-1}(a^*)$. In fact, j is antilinear since $*$ is; j is multiplicative since both i^{-1} and $*$ are antimultiplicative, and j reverses comultiplication since i^{-1} does so while $*$ conserves it. Finally,

$$a \xrightarrow{j} i^{-1}(a^*) \xrightarrow{j} i^{-1}[i^{-1}(a^*)^*] = i^{-1} \circ i(a) = a \text{ (in view of d)),}$$

$$a \xrightarrow{ij} a^* \xrightarrow{ij} a .$$

Conversely, let H be a Hopf algebra with $*$–structure j. Put $a^* = ij(a)$ for all $a \in H$. Clearly, this has all the algebraic properties of a $*$–involution. However H is not normed. In [W2], S.L. Woronowicz suggests the following construction. Consider the class of all $*$–representations π of H in Hilbert spaces. Define $||a|| = \sup||\pi(a)||$ and consider a completion $H \to \overline{H}$ with respect to this seminorm. If this is an injection (as in the case for $H = GL_q(2)$), we get in this way a compact matrix pseudogroup, provided H is generated by the entries of a multiplicative matrix Z.

We strongly recommend to the reader [W1], [W2] and subsequent papers by Woronowicz who has actually laid down the fundamentals of the representation theory of compact matrix pseudogroups, i.e. the "compact forms" of our quantum groups.

11. YANG -BAXTER EQUATIONS

1. Yang–Baxter operator. Let F be a linear space, $R : F\otimes F \to F\otimes F$ an invertible linear map. It is well known that if $R = S_{(12)} : f_1 \otimes f_2 \to f_2\otimes f_1$, then one can define a representation of the symmetric group S_n on $F^{\otimes n}$ by the following prescription : represent a substitution $\sigma \in S_n$ as a product of transpositions of neighbours and apply $R_{i,i+1} = S_{(i,\ i+1)}$ instead of each $(i, i+1)$. Of course, such a decomposition of σ is non–unique but the resulting linear operator does not depend on it.

An arbitrary operator R is called *a* **Yang–Baxter operator if it** shares this property with $S_{(12)}$, i.e. it defines a series of representations of S_n on $F^{\otimes n}$ by the construction described above.

2. Proposition. R is a Yang–Baxter operator iff it satisfies the following Yang–Baxter (or triangle) equations :

$$R_{23}R_{12}R_{23} = R_{12}R_{23}R_{12} : \ F^{\otimes 3} \to F^{\otimes 3} \qquad (1)$$

$$R_{12}^{2} = \text{identity} : \ F^{\otimes 2} \to F^{\otimes 2} \qquad (2)$$

where, say $R_{23}(f_1\otimes f_2\otimes f_3) = f_1 \otimes R(f_2\otimes f_3)$ etc.

Proof. This follows directly from the fact that the Coxeter relations among the transpositions define the symmetric group. ■

3. Examples.

a) $F = F_0\oplus F_1$, where $\{0, 1\} = Z_2$. Set $\tilde{f} = 0$ (resp. 1) if $f\in F_0$ (resp. $f \in F_1$). Put

$$R(f\otimes g) = (-1)^{\tilde{f}\tilde{g}} g\otimes f.$$

This Yang–Baxter operator in a very definite sense "generates" the main notions of superalgebra (cf. below).

b) More generally, let $F = \oplus F_\alpha$ where α runs over a set of indices.

Define

$$R(f\otimes g) = a_{\alpha\beta}\, g\otimes f \quad \text{if} \quad f\in F_\alpha,\ g\in F_\beta;\ a_{\alpha\beta}\in k\,.$$

Then (1) is satisfied automatically while (2) means that

$$a_{\alpha\beta}\, a_{\beta\alpha} = 1 \quad \text{for all} \quad \alpha, \beta.$$

c) Let H be a Hopf algebra and F_1, F_2 be H–comodules given by $\delta_i : F_i \to H\otimes F_i$. We can define on $F_1\otimes F_2$ the structure of an H–comodule by means of the following composed map :

$$F_1\otimes F_2 \xrightarrow{\delta_1\otimes\delta_2} H\otimes F_1\otimes H\otimes F_2 \xrightarrow{S_{(23)}} H\otimes H\otimes F_1\otimes$$

$$\otimes F_2 \xrightarrow{m\otimes id} H\otimes F_1\otimes F_2\ .$$

We recommend to the reader to check the axioms of §2.8 and to calculate the corresponding multiplicative matrix (cf. Proposition 2.9).

Of course, $F_2\otimes F_1$ is also an H–comodule so that $F\otimes F$ has <u>two</u> natural structures of a comodule. However, they may be non–isomorphic!

If we apply this construction to the "universal" case $F = H$ and ask when the two natural corepresentations on $H\otimes H$ are equivalent we can investigate the case when this equivalence is established by means of a conjugation operator :

$$R : H\otimes H \to H\otimes H, \quad R(x) = \rho x \rho^{-1},\ \rho\in H\otimes H.$$

We leave to the reader the task of rewriting (1) and (2) : compare [L] and [FRT], cf. also the discussion of the "triangle" Hopf algebras in [D1].

4. Quadratic algebras generated by a Yang–Baxter operator. Let $R \in \mathrm{End}(F \otimes F)$ be a Yang–Baxter operator. Then we can form an "R–symmetric algebra" of F :

$$S_R^\bullet(F) = \bigoplus_{n \geq 0} \{S_n\text{–invariants of } F^{\otimes n}\},$$

and an "R–exterior algebra of F :

$$\Lambda_R^\bullet(F) = \bigoplus_{n \geq 0} \{f \in F^{\otimes n} \mid \sigma(f) = \mathrm{sgn}(\sigma) f\}.$$

One easily checks that they are in fact quadratic :

$$S_R^\bullet(F) \longleftrightarrow \{F, \ (\mathrm{Id}-R)(F \otimes F)\},$$

$$\Lambda_R^\bullet(F) \longleftrightarrow \{F, \ (\mathrm{Id}+R)(F \otimes F)\}$$

in the notation of §2. For $S = S_{(12)}$, we obtain the usual polynomial and exterior algebras. Note that $S_R^\bullet(F)^! \cong \Lambda_{R^*}^\bullet(F^*)$.

Starting with such an algebra A, we can generate $\underline{end}(A)$, $\underline{end}(A, g)$ and their Hopf envelopes $\underline{g\ell}(A)$, $\underline{g\ell}(A, g)$.

In particular, for the space with a marked basis

$$F = \bigoplus_{i=1}^{n} k x_i, \quad R(x_i \otimes x_j) = q^{j-i} x_j \otimes x_i \tag{3}$$

we can repeat word for word everything we did in §1. Namely, $S_R^\bullet(F)$ defines a "quantized" n-space $A_q^{n|0}$, $\Lambda_R^\bullet(F)$ defines a similar superspace

$A_q^{0|n}$ which is grassmannian (see §8.1) and hence allows us to define a quantum determinant. In this way we obtain the quantum groups $GL_q(n)$, $SL_q(n)$. Supplying a $*$–structure as in §10 we can get Woronowicz's $U_q(x)$.

Of course, we are not obliged to restrict ourselves to a one–parameter deformation of $GL(n)$. As in 3b), we can consider a more general Yang–Baxter operator

$$F = \bigoplus_{i=1}^{n} kx_i, \quad R(x_i \otimes x_j) = a_{ij} x_j \otimes x_i, \quad a_{ij} a_{ji} = 1 \tag{4}$$

and again repeat the constructions of §1. A warning : if all $a_{ii} = 1$, $\Lambda_R^{\bullet}(F)$ will still be grassmannian. However, if some $a_{ii} = 1$ and some $a_{ii} = -1$, we get a kind of intrinsic Z_2–grading upon F and our quantum $GL_R(F)$ will have some resemblance to the general linear supergroup $GL(r|s)$. In order to understand more clearly what this means, let us digress a bit and discuss the role of the transposition operators S_σ, $\sigma \in S_n$ in our principal constructions.

5. Transposition operators. We invite the reader to turn to §2 and to look through the diagrams defining bialgebras, Hopf algebras and various constructions on them. We see that $S_{(12)}$ appears in at least five places explicitly :

a) In the connecting axiom (§2.1)

b) In the definition of m^{op}, Δ^{op} (§2.2).

c) In the Theorem 2.4 on the uniqueness of the antipode.

Moreover, we omitted in §2 several checks and proofs referring the reader to Abe[2]. If one looks carefully at which properties of $S_{(12)}$ are used in these proofs one can verify that they are precisely the Yang–Baxter equations (1), (2)!

Therefore, having chosen a Yang–Baxter operator $R \in \mathrm{End}(F \otimes F)$, one can define a whole series of "R–relativized" notions by just replacing $S_{(12)}$ by R in all definitions and proofs. Here is a list of possibilities for the reader to get acquainted with this idea :

- define a structure of R–Hopf algebra on F;
- translate the main constructions of §3 to quadratic algebras {F, R(F)} using R instead of $S_{(12)}$;
- construct $\underline{end}_R(A)$ for an R–quadratic algebra A and prove that it is an R–bialgebra.

In particular, making everything Z_2–graded and using the "sign rule" of Example 3.a one gets a superization of our class of quantum groups (and quantum spaces).

Of course, in general it is rather awkward to have only one Yang–Baxter operator acting upon space $F \otimes F$. One should be able to transpose different linear spaces, to form tensor products of Yang–Baxter operators, to dualize, etc. There is a very convenient axiomatic framework for all this; namely, that of tensor categories. In the next section we shall briefly review it.

6. Weak Yang–Baxter operators and braid groups. An invertible element $R \in \mathrm{End}(F \otimes F)$ satisfying (1) but not necessarily (2) is called a weak Yang–Baxter operator.

Having a weak Yang–Baxter operator, we can define an action of the Artin braid group Bd_n upon each tensor product $F^{\otimes n}$ in such a way that $R_{(i,i+1)}$ corresponds to a standard generator τ_i of the braid group (say, the i–th strand goes over the (i+1)–th one). In fact, here is the graphic representation of (1) in terms of braids :

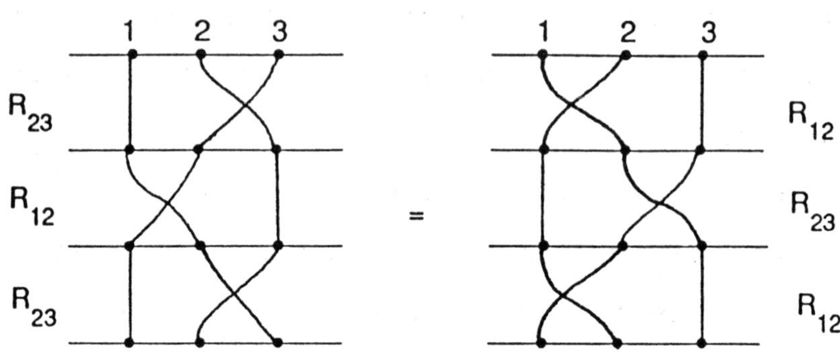

Weak Yang–Baxter operators were recently extensively used in knot theory and related domains : cf. papers by Jones, Turaev, Fröhlich.

However, I cannot connect them directly with quantum groups. Of course, formally one can still repeat the constructions of §4, but the resulting algebras will probably be too small : cf. Proposition 6.8. The correct way to enlarge them is described in §6.9; one should use the matrix relations (2), §6 : $R(Z \odot Z) = (Z \odot Z)R$, which are only "a part" of the relations defining $\underline{end}(S^{\bullet}_R(F))$, if $R^2 \neq 1$.

7. Summary. Quantum groups are related to Yang–Baxter operators in two ways.

First, a Yang–Baxter operator defines a specific quantum space whose quantum symmetry groups can be constructed in the way described above.

Second, a Yang–Baxter operator can be entered in the very definitions of all fundamental objects, thus changing them drastically. The best way to describe this systematically is to start with a collection of, say, linear spaces, stable with respect to tensor product and dualization and endowed with a collection of Yang–Baxter operators $V \otimes W \to W \otimes V$ defined for each pair of this collection of spaces.

In short, we start with an abstract "tensor category" and repeat everything in it. We get a new theory with the same logical structure.

In the next section we describe it in more detail.

12. ALGEBRA IN TENSOR CATEGORIES AND YANG–BAXTER FUNCTORS

1. Tensor categories: a summary. In this subsection we reproduce in a very condensed manner the axiomatization of a class of categories which are sufficiently similar to the category of vector spaces over a field k to allow us to reproduce most of the previous constructions in an abstract setting. See [DM] for details and [B] for more references and some important constructions.

Let S be a category and

$$\otimes : S\times S \to S, \quad (X,Y) \to X\otimes Y$$

be a functor. An associativity constraint for $(S,\otimes)$ is a functorial isomorphism

$$\varphi_{X,Y,Z} : X\otimes(Y\otimes Z) \xrightarrow{\sim} (X\otimes Y)\otimes Z$$

verifying a "pentagon axiom" which implies that one can write a tensor product of an arbitrary finite family of objects without bothering about brackets. A commutativity constraint is a functorial isomorphism

$$\psi_{X,Y} : X\otimes Y \xrightarrow{\sim} Y\otimes X$$

satisfying a compatibility condition with φ which implies that for any family of objects $X_1,\dots,X_n$ and $\sigma \in S_n$ one can define an isomorphism $S_\sigma : X_1\otimes\dots\otimes X_n \xrightarrow{\sim} X_{\sigma^{-1}(1)}\otimes\dots\otimes X_{\sigma^{-1}(n)}$ in such a way that $S\sigma S\tau = S\sigma\tau$. Of course, S_σ are composed from elementary transpositions ψ_{X_i,X_j}. An identity object U of S is in fact a pair $(U,\ u : U \xrightarrow{\sim} U\otimes U)$ such that $X \to U\otimes X : S \to S$ naturally extends to an equivalence of categories.

A <u>tensor category</u> is a category S together with a tensor product functor $\otimes$, compatible constraints of associativity and commutativity, and a unit object.

Examples.

a) Vector spaces over a field with

$$\varphi(x\otimes(y\otimes z)) = (x\otimes y)\otimes z; \quad \psi(x\otimes y) = y\otimes x; \quad x\in X,\ y\in Y,\ z\in Z$$

b) Z_2–graded vector spaces over a field with the same φ and $\psi(x\otimes y) = (-1)^{\tilde{x}\tilde{y}} y\otimes x$ where $\tilde{x}$ is the parity of x.

c) The category of quadratic algebras QA with either $\circ$, or $\bullet$ as the tensor product (cf. §3).

d) A category of representations of an affine group scheme (cf. [DM]).

e) A category of (co)representations of a Hopf algebra with an additional structure required to define φ and ψ satisfying the necessary axioms (cf. [L]).

2. Rigid tensor categories. In a rigid tensor category one can define for any pair of objects X, Y their internal <u>Hom</u> object $\underline{\mathrm{Hom}}(X, Y)$ together with a functorial isomorphism

$$\underline{\mathrm{Hom}}(Z, \underline{\mathrm{Hom}}(X,Y)) \xrightarrow{\sim} \underline{\mathrm{Hom}}(Z\otimes X, Y)$$

and functorial isomorphisms

$$\bigotimes_{i\in I} \underline{\mathrm{Hom}}(X_i, Y_i) \to \underline{\mathrm{Hom}}(\bigotimes_{i\in I} X_i, \bigotimes_{i\in I} Y_i)$$

(cf. [DM]). All examples of the previous subsection (with appropriate finiteness conditions) are rigid, with the exception of (QA, $\circ$) where there are no internal Hom's.

In particular, in a rigid category there is a dualization functor $X \to X^{\vee} = \underline{Hom}(X, U)$ where U is a unit object, and all objects are reflexive : $X^{\vee\vee} \cong X$.

3. Abelian tensor category. We shall call a rigid tensor category S abelian if it is an abelian category and $\otimes$ is biadditive. We usually also assume that $Hom(X, Y)$ are linear spaces over a field k, i.e. that S is a k–category. The tensor product is then an exact functor.

4. Yang–Baxter functors. Let $(S, \otimes)$ and $(T, \otimes)$ be two tensor categories. A faithful functor $F : S \to T$ is called a Yang–Baxter functor if it transforms the tensor product in S into the tensor product in T and is compatible with the associativity constraints in S, T but not necessarily with the commutativity constraints.

Speaking somewhat loosely, one can imagine an object of S as an object of T endowed with an additional structure in such a way that the commutativity constraint in S differs from that in T in a way that takes into account this additional structure.

Example : S = Z_2–graded linear spaces, T = linear spaces, F = forgetful functor.

Lyubashenko [L] describes how to construct a rigid tensor category $(S, \otimes)$ with a Yang–Baxter functor to vector spaces starting from just one Yang–Baxter operator verifying a certain non–degeneracy assumption. Essentially one adds all tensor products, duals, subspaces and quotient spaces "compatible" with the original Yang–Baxter operator so that it generates the new commutativity constraint on the obtained category. (Of course, morphisms are also restricted by a compatibility condition). Lyubashenko has also proved a theorem showing that such Yang–Baxter functors are essentially forgetful functors from a category of representations of a Hopf algebra.

5. Algebra in abelian tensor categories. We shall now fix an abelian tensor category $(S, \otimes)$ and explain how one can imitate the standard algebraic constructions in order to define the relativized notions of an S–algebra, S–coalgebra, S–Lie–algebra, S–quantum group etc. A word of warning is in order. Our axioms are modelled upon the properties of finite–dimensional vector spaces. This is essential everywhere when one needs the reflexivity properties. On the other hand, a symmetric algebra of a vector space is infinite–dimensional, etc. Therefore we must in fact extend our initial universe $(S, \otimes)$ so as to include infinite direct sums, projective and injective limits, etc. We shall pretend that all this has somehow been done and forget about it (cf. [B] for a version of the solution of this important problem).

a) S–algebra. An associative S–algebra is an object A in S together with a multiplication morphism m : $A\otimes A \to A$ verifying the associativity diagram of §2.1. Similarly, an associative S–algebra with a unit consists of A, m and a morphism $\eta : U \to A$, verifying the unit diagram of §2.1. Here U is a unit object of S.

It is clear now how to define an S–algebra, S–bialgebra, S–Hopf algebra imitating §2. The notion of an S–Lie–algebra is probably less evident. It is convenient to start with the definition of an S–commutator.

b) $\underline{S\text{-commutator}}$. Let (A, m) be an associative S–algebra. Define the commutator map :

$$[\bullet]_S : A\otimes A \to A, \quad [f]_S = m(f) - mS_{(12)}(f)$$

(we return to the S_σ–notation instead of $\psi_{X,Y}$–notation). For S = {vector spaces} we get the usual formula

$$[a\otimes b]_S = ab - ba.$$

One easily checks the following identities :

$$\text{"S–anticommutativity"} : [S_{12}(f)]_S = -[f]_S \qquad (1)$$

$$\text{"S–Jacobi identity"} : \{g\}_S + \{S_{(312)}(g)\}_S + \{S_{(231)}(g)\}_S = 0 \qquad (2)$$

where

$$\{\bullet\}_S : A\otimes A\otimes A \xrightarrow{id\otimes[\bullet]_S} A\otimes A \xrightarrow{[\bullet]_S} A \qquad (3)$$

is the triple commutator $[a, [b, c]]$.

We can now define S–commutative algebras as those for which $[\bullet]_S = 0$ identically. Rewriting this condition in terms of a commutative diagram, we can also reverse arrows and define S–cocommutative S–algebras, etc. It is instructive to look at where the definitions of $A_S^!$, A_S^{op}, $A \otimes_S B$ etc. differ from those of $A^!$, A^{op}, $A\otimes B$, etc.

c) <u>$\mathcal{S}$-Lie-algebras</u>. An $\mathcal{S}$–Lie–algebra is an object L of $\mathcal{S}$ together with a morphism $[\bullet] : L \otimes L \to L$ verifying (1) and (2) (note that the triple commutator (3) is defined in terms of $[\bullet]$ only).

d) <u>Differential operators</u>. Let A be an associative $\mathcal{S}$–algebra.

Following Grothendieck, we can define an internal object of $\mathcal{S}$–differential operators $\mathrm{Diff}_{\leq i}(A)$ of order $\leq i$ together with the left action map

$$\ell_i : \mathrm{Diff}_{\leq i}(A) \otimes A \to A$$

by the following inductive procedure : $\mathrm{Diff}_0(A) = A$, $\ell_0 = m$. In general, $\mathrm{Diff}_{i+1}(A)$ is the maximal subobject of $\underline{\mathrm{Hom}}(A, A)$ (which is an $\mathcal{S}$–algebra) such that

$$[\mathrm{Diff}_{\leq i+1}(A), \mathrm{Diff}_0(A)]_{\mathcal{S}} \subset \mathrm{Diff}_{\leq i}(A)$$

and ℓ_{i+1} is induced by the canonical map $\underline{\mathrm{Hom}}(A, A) \otimes A \to A$. Of course, the existence of such a subobject is a separate matter. Having $\mathrm{Diff}_{\leq i}$, one can formally introduce quantum jet spaces.

e) <u>Quadratic $\mathcal{S}$-algebras</u>. The constructions of §§3, 4 can be repeated word for word.

We hope that the attentive reader has by now grasped the rules of the game. One of the interesting outcomes is a rich new supply of complexes defining $\mathcal{S}$–relativized versions of Koszul algebras, Hochschild (co)homology, cyclic (co)homology etc.

6. A proposal for a non–commutative algebraic geometry universe. Take an abelian tensor k–category $(\mathcal{S}, \otimes)$ such that there exists a Yang–Baxter functor $(\mathcal{S}, \otimes) \to \{\text{k–vector spaces}\}$. Consider the category of *S*–commutative graded algebras A and try to initiate the Serre–style projective algebraic geometry in this category. In particular, such an algebra A defines in an intrinsic way the abelian category "Coh Proj A" of "coherent sheaves on Proj A" : just divide graded *S*–A–modules by finite ones. Morphisms between such projective spectra should be defined as certain functors among "Coh Proj A" ("inverse images"). A realization of this program has been started by A. Verevkin and will appear in forthcoming papers.

In short, as A. Grothendieck taught us, to do geometry you really don't need a space, all you need is a category of sheaves on this would–be space.

13. SOME OPEN PROBLEMS

In conclusion we state here in a rather unsystematic way some questions and indicate directions for further research related to the approach to quantum groups described in these lectures.

1. Differential geometry. One should be able to define the notions of the de Rham complex, tangent and cotangent spaces etc. for a quantum quadratic space and its <u>end</u>–bialgebra and then to transport them to the corresponding quantum group. There is no doubt that it can be done in a standard way in the category of quadratic *S*–commutative algebras, where *S* is a Yang–Baxter category of linear spaces. The Woronowicz differential forms (see [W2]) on SU(2) seem to fit into this framework.

Is it possible to broaden significantly the scope of this differential geometry? The Koszul complex $K^*(A)$ vaguely looks like a complex of noncommutative currents (not forms, due to an additional dualization $A^! \to A^{!*}$).

Of course, we have Alain Connes' suggestion : to use cyclic cohomology.

2. Cyclic (co)homology. For quadratic algebras A, this was studied by Feigin and Tsygan. One should understand whether the "quantum symmetries" of A act upon its Hochschild and cyclic (co)homology and study the cyclic (co)homology of quantum (semi)groups.

3. Root technique and Kac–Moody quantum groups. Chevalley, Springer and others have applied the classical root technique directly to the function ring of an algebraic group instead of its Lie algebra. Can one extend this to our quantum groups, perhaps by first quantizing the fundamental representation treated as a "quadratic quantum space"? Can one then use some (version of) Kac–Moody Cartan matrices?

4. Quantum Virasoro? The same question for the Virasoro "group". One possible approach is to use the embedding of the Virasoro algebra into the (central extension of) an infinite dimensional linear Lie algebra. Since we have a rich supply of quantum linear groups and some of our constructions modify easily to define various kinds of $GL_{quant}(\infty)$, we may hope also to quantize Virasoro. The new effects of infinite dimensions should be studied carefully.

5. Flag spaces, quantum fibrations, Yang–Mills. It would be very important to define noncommutative flag spaces for quantum groups or at least principal fibrations over them. This chapter of noncommutative (algebraic) geometry is wide open. Of course, one hopes that for "good" groups (like the symmetries of *S*–symmetric algebras) one should obtain Schubert cells, Borel–Weil–Bott theory etc. Here, one should not act too hastily since even in supergeometry this program was started only recently and revealed both rich content and some puzzling new phenomena.

6. Representations of rings of functions. The initial source of quantum group theory, the quantum inverse transform method, requires the study of (unitary) representations of function rings of quantum groups.

See also articles by Woronowicz, and Vaksman–Soibelman.

7. Koszul rings and differential graded algebras. The class of quadratic Koszul algebras is a very important one. Among its many interesting properties one should mention that if A is Koszul than A! is canonically isomorphic to $\mathrm{Ext}^*_A(k,k)$. (Otherwise it is a subalgebra generated by Ext^1). This makes one wonder whether it is possible to generalize our constructions of quadratic algebras, extending them to a certain category of differential graded algebras (with morphisms considered up to homotopy).

8. "Hidden symmetry" in algebraic geometry. An arbitrary projective algebraic manifold (actually, even a scheme) is a projective spectrum of a quadratic algebra. For some very important manifolds like abelian varieties (with rigidity) and their moduli spaces such quadratic algebras come equipped with canonical generators of degree one (Mumford's theory of abstract theta – functions). Our results show that certain universal Hopf algebras act upon such manifolds. Their properties deserve closer investigation.

Probably, a simpler class of examples is furnished by canonical embeddings of algebraic curves. With some rare exceptions, a curve of genus g embeds into P^{g-1} with the help of its differentials of the first kind, and all relations among them are generated by quadratic ones. The corresponding Hopf algebras of hidden symmetries of a curve may in principle be much larger that classical automorphism group algebras (which are finite–dimensional for $g \geq 2$ and usually trivial).

BIBLIOGRAPHY

[A] E. Abe, Hopf algebras, Cambridge Tracts in Math., vol. 74, Cambridge Univ. Press, 1980.

[B] M. Barr, *-Autonomous categories, Lecture Notes in Math, 752, Springer-Verlag, Berlin-Leidelberg-New York, 1979.

[BF] J. Backelin, R. Fröberg, Koszul algebras, Veronese subrings and rings with linear resolutions. Revue Roumaine de Math. Pures et Appl., t. xxx, no.2 (1985), 85-97.

[D1] V.G. Drinfeld, Quantum groups, Proc. Int. Congr. Math. Berkeley, 1986, vol. 1, 798-820.

[D2] V.G. Drinfeld, On quadratic commutation relations in the quasiclassic limit, In: Mat. Fizika is Funkc. Analiz. Kiev. Naukova Dumka (1986), 25-33 (in Russian).

[DM] P. Deligne, J. Milne, Tannakian categories, Lecture Notes in Math., vol. 900, Springer-Verlag, Berlin-Heidelberg-New York, 1982, 101-228.

[F] L.D. Faddeev, Integrable models in (1+1)-dimensional quantum field theory (Lectures in Les Houches, 1982), Elsevier Science Publishers B.V., 1984.

[FRT] L.D. Faddeev, N.Y. Reshetikhin, L.A. Takhtajan, Quantization of Lie groups and Lie algebras, preprint LOMI (1987).

[FT] L.D. Faddeev, L.A. Takhtajan, Hamiltonian approach to solitons theory, Springer Verlag, Berlin–New York, 1987.

[J1] M. Jimbo, A q–difference analogue of U(g) and the Yang–Baxter equation, Lett. Math. Phys. 11 (1986), 247-252.

[J2] M. Jimbo, A q–analogue of U(gℓ(N+1)), Hecke algebra and the Yang–Baxter equation, Lett. Math. Phys. 11 (1986), 297-252.

[L] V.V. Lyubashenko, Hopf algebras and vector symmetries, Uspekhi Math. Nauk 41 (1986), no. 5, 185-186 (in Russian).

[Lö] C. Löfwall, On the subalgebra generated by one–dimensional elements in the Yoneda Ext–algebra, Lecture Notes in Math., vol. 1183 (1986), Springer, Berlin–New York, 291-338.

[M] Y.I. Manin, Some remarks on Koszul algebras and quantum groups, Ann. Inst. Fourier, Tome XXXVII, f. 4 (1982), 191-205.

[P] S.B. Priddy, Koszul resolutions, Trans AMS, 152: 1 (1970), 39-60.

[S] E.K. Sklyanin, The quantum version of the inverse scattering method, Zap. Nauchn. Sem. LOMI 95 (1980), 55–128 (in Russian).

[S] Ya. S. Soibelman, Irreducible representations of the functional algebra of quantized SU(n) and the Schubert cells, preprint, 1988.

[STF] E.Sklyanin, L. Takhtajan, L. Faddeev, Quantum inverse scattering transform method, TMF, 40, no. 2 (1979) 194-220 (in Russian).

[VS] L.L. Vaksman, Ya. S. Soibelman, The algebra of functions on quantizied SU(2), Funkc. Analiz i ego Priloz (in Russian).

[W1] S.L. Woronowicz, Twisted SU(2)-group, An example of a non-commutative differential calculus, Publ. RIMS, Kyoto Univ., vol. 23, No. 1 (1987), 117-181.

[W2] S.L. Woronowicz, Compact matrix pseudogroups, Comm. Math. Phys. 111 (1987), 613–665.